Certain Extensor Structures in the Calculus of Variations

Certain Extensor Structures in the Calculus of Variations

William Clifton Bean

Senior Engineer, NASA–Houston (ret)
Instructor of Mathematics, University Of Houston–Clear Lake
50-year member, Tensor Society

YBK Publishers New York

Certain Extensor Structures in the Calculus of Variations

YBK Publishers, Inc.
39 Crosby Street
New York, NY 10013
www.ybkpublishers.com

ISBN: 978-1-936411-40-5

Library of Congress Control Number: 2015949971

Manufactured in the United States of America for distribution in
North and South America or in the United Kingdom or Australia
when distributed elsewhere.

For more information, visit
www.ybkpublishers.com

Foreword

In 1971, I completed the work that follows as a doctoral dissertation entitled, *On Certain Extensor Structures in the Calculus of Variations*. The document was the culmination of research begun in 1959 in the mathematics department of the University of Texas-Austin under the supervision of Professor Homer V. Craig, to whom this release is dedicated.

I offer my dissertation for publication as a book for several reasons:

(1) There is a certain gain in coherency by publishing its interrelated topics as a monograph.
(2) Much of the work consists of unpublished original research that Dr. Craig estimated could be converted into at least four technical papers. Unless this material is developed independently by others, there is a significant risk that, if not published for wider distribution, much of it will be lost.
(3) The introduction (chapter I) provides a thorough review of many of the fundamentals of extensor theory. This could provide the impetus to study extensor methodology for later generations who cannot easily obtain access to Dr. Craig's 1943 McGraw Hill textbook, *Vector and Tensor Analysis*, now out of print.
(4) Lastly, redistribution of the dissertation is intended to honor Professor Homer V. Craig. In guiding the preparation of a lengthy dissertation in the calculus of variations for many years, he undoubtedly diverted much of the energy he might otherwise have devoted to writing his own book about the calculus of variations. It was a book he had always wanted to write if the opportunity arose.

In order to trace the history of research involving extensors, in the process of placing my work within its proper context, I recently did a literature search, focusing primarily on the issues of Tensor Magazine, New Series (subsequently referred to simply as Tensor) since its inception as an international journal in 1950. Henceforth, bold face volume numbers refer to the journal Tensor except where explicitly stated otherwise.

Dr. Craig was strongly motivated by his belief in invariance as a dominant notion in mathematics and physics. His quest for sets of quantities whose transformation properties involve invariance led him to introduce the concept of extensor in his 1937 paper "On Tensors Relative to the Extended Point Transformation." That paper appeared in volume

59 of the American Journal of Mathematics. Therein, Dr. Craig noted that extensors have properties for which there is no tensor analog, and went on to show that certain quantities due to Zermelo are extensor components. Dr. Craig developed extensors further in his 1943 textbook, *Vector and Tensor Analysis*. He went on to write numerous papers about applications of extensors, and about extensions of the extensor concept, during the course of his academic career. His goal was to unify, by means of extensor theory, a variety of special topics taken mostly from mathematical physics and the calculus of variations. Where the base functions were not presented explicitly in the problems themselves, he frequently took them to be those functions involved in a conservation assumption.

Dr. Craig applied extensors to such topics as the Lagrangian equations of motion in Mathematics Magazine **26** (1949), the theory of partial differential equations and Poisson brackets in **6** (1956), a nonconservative dynamical system in **10** (1960), and an extended phase space in Annali di Matematica, Series IV, **58** (1962). Regarding his early work in the calculus of variations, he co-authored with C. W. Horton a paper on extensors and the Hamiltonian equations in **1** (1951). He laid the groundwork for further applications of extensors to the calculus of variations when he wrote about linear extensor equations in **4** (1954) and **5** (1955). In those two papers Dr. Craig emphasized (1) choosing numerical coefficients so as to avoid overdetermination, and (2) applying displacement as needed to secure alignment of physical units.

Another giant in differential geometry was A. Kawaguchi, who wrote at least four technical papers concerning extensors during 1938-1940. A. Kawaguchi introduced the idea of multiple-parameter extensors in 1939, and investigated them at some length in a seminal paper in the Journal of the Faculty of Science of Hokkaido Imperial University **9** (1940). In that same 1940 paper, A. Kawaguchi also proved the extensor character of certain quantities which are extensive derivatives.

During his prolific career, A. Kawaguchi produced a vast body of research, some of it involving extensors. He "published 143 original papers in mathematics" and "36 books, monographs, and textbooks," according to R. S. Ingarden, writing in **45** (1987), a commemorative volume of Tensor. A major contribution of A. Kawaguchi was the discovery, circa 1932-1937, of the geometry of Kawaguchi spaces, so dubbed by J. L. Synge in the American Journal of Mathematics **57** (1935).

To emphasize the fertility of extensor research, and to exploit the literature search, I shall wish to close by tracing the evolution of that research. The intent is not to furnish an exhaustive summary, but rather to offer a sufficiently detailed overview to exemplify several lines of inquiry involving extensors. Naturally, I wish to begin with the line of inquiry that features the calculus of variations, paying particular attention to how my work fit into the evolving picture.

Though the literature search revealed that extensor analysis was indeed a fertile area for research from the 1940's through the 1990's, it exposed a relative paucity of work not done by Craig that linked extensors and the calculus of variations. Closely related work included S. J. Aldersley's "Higher Euler operators and some of their applications" in the Journal of Mathematical Physics **20** (1979), Robert H. Bowman's "Primary Extensors and The Bolza Problem," **34** (1980), Yoshihide Watanabe's "On the Extensions of the Euler Operator in the Calculus of Variations," **41** (1984), and several papers by Mimura and Nono in **48** (1989) and **49** (1990) on an inverse problem of Lagrangian dynamics- namely, for a first-order Lagrangian function, to find necessary and sufficient conditions for a given system of differential equations to be identified with the Euler-Lagrange equations. Another related paper was Shi-Wei Niu's "Higher order variational calculus," **57** (1996), which discussed the Hamilton-Jacobi equation and its solution by virtue of the higher order jet bundle and extended phase space. Finally, we should note that A. Kawaguchi gave a paper to the Greek Mathematical Society in September 1973 on the geometrization of the calculus of variations.

In 1959 I arrived as a graduate student under Craig, just after he contributed two more key papers in the calculus of variations, namely (1) "On Extensors in the Calculus of Variations," Mathematics Magazine **30** (1957), and (2) "On Primary Extensors" **8** (1958). He followed those papers with another about physical applications of the gradient and Kawaguchi extensors in **13** (1963). In the 1957 and 1958 papers, Craig showed that the Euler equations for a simple calculus of variations of the type $\int_{t1}^{t2} F(x, x')\,dt$ can be expressed as equality of the two primary extensors associated with F (x, x'), namely the gradient extensor F;αa and a Kawaguchi extensor Fαa constructed from the tensor rank of F;αa.

One evening in 1959, I was working at the blackboard in Dr. Craig's graduate course on research in tensor analysis. Following Dr. Craig's suggestion, I used a differencing technique to discover that the Euler equations associated with a simple calculus of variations problem of the type $\int_{t1}^{t2} F(x, x', x'')\,dt$ could be expressed as a linear extensor equation involving a gradient extensor, a Kawaguchi extensor, and one other extensor. It seemed to me that this important discovery should readily lend itself to generalization. Accordingly, I spent the following weekend deriving a formula that generalized the concept of primary extensor so as to apply to a class c^{M+1} invariant function $F(x, x', \ldots, x^{(M)})$ evaluated over a class c^{2M} parameterized arc. This involved constructing from $F(x, x', \ldots, x^{(M)})$ not only (1) a gradient extensor 0αa (identical to F;αa), but also (2) a set of M additional extensors Lαa (with L = 1, 2, ..., M) whose ranks involve terms composed of parameter derivatives of the quantities F;0a, F;1a, ..., F;Ma and whose zero ranks are given by $L0a = F;La^{(L)}$ for L = 0, 1, ..., M. Here, Mαa is the Kawaguchi extensor. The quantities Lαa for each L = 0, 1, ..., M are defined according to (12.8) in the 1971 dissertation (herein reissued), but were described first in my August 1960 master's thesis entitled *Higher Order Euler Equations as Linear Extensor Equations.*

Dr. Craig was quite taken with this result. One of the first things he had me do was prove that my formula generated quantities which were indeed extensors. My derivation method had established a kind of necessity, but not sufficiency. I proceeded to do this, although at the time I did not find the most straightforward proof. It would have been sufficient to observe that each Lαa is a linear combination of other quantities which are themselves extensors *by construction*. Therefore, each Lαa is an extensor *by construction*.

Eventually, we were able to show that each of the M extensors Lαa, L = 1, 2, ..., M, forms a perfect Lth derivative when contracted with $V^{a(\alpha)}$, where V^a is an arbitrary class c^M tensor. Accordingly, it was noted that the linear extensor equation

$$0\alpha a - 1\alpha a + 2\alpha a - \ldots + (-1)^M M\alpha a = 0$$

converts the first variation to the integral of a derivative, so that the first variation vanishes. Each nonzero rank of the above equation vanishes. This promotes rank 0 to tensor rank, and rank 0 expresses the Euler equations associated with $F(x, x', \ldots, x^{(M)})$, namely

$$F;0a - F;1a' + F;2a'' - \ldots + (-1)^M F;\, Ma^{(M)} = 0.$$

Next, Dr. Craig suggested that I work on the isoperimetric problem of the type $\int_{t1}^{t2} F(x, x', \ldots, x^{(M)})\, dt$, $\int_{t1}^{t2} G(x, x', \ldots, x^{(M)})\, dt = \text{const}$. It developed that the appropriate Euler equations could be expressed as a linear equation relating the gradient extensors F;αa, G;αa and 2M additional extensors whose zero ranks, respectively, were F;1a', ... $F;Ma^{(M)}$ and G;1a', ..., $G;\, Ma^{(M)}$. In other work, we found an extensor solution to a surface integral problem of type $\iint_{D0} F(x, x^{(0,1)}, x^{(1,0)}, \ldots, x^{(M1,\, M2)})\, du_1\, du_2$. Much of this work was reported in the 1960 master's thesis.

Between 1963 and 1970, while I was dividing time between my research and my work at NASA-Houston, there were a couple of publications that influenced somewhat the development of my dissertation. (1) First, in September 1966, Dr. Craig published "An Extensor Generalization of a Simple Calculus of Variations Problem" in volume 17 of Tensor. Therein, he showed that the vanishing of the Hamiltonian tensor associated with the extensor Eαa is a necessary and sufficient condition that the arc C_0 is a normalizing arc in a modified calculus of variations problem in which, rather than the vanishing of the first variation of an integral, there is investigated ab initio the requirement $I_1 = 0$, where I_1 is defined as $\int_{t1}^{t2} E\alpha a\, V^{a(\alpha)}\, dt$. Here, Eαa is an arbitrary excovariant extensor of range α: 0 to M, and V^a is a variation tensor which generates a family of admissible comparison arcs. Thus the concept of a normalizing arc with respect to the extensor Eαa as an arc along which I_1 vanishes replaces the concept of a stationary arc as an arc along which the first variation $\int_{t1}^{t2} F;\alpha a\ V^{a(\alpha)}\ dt$ of $\int_{t1}^{t2} F\ (x, x', \ldots, x^{(M)})\ dt$ vanishes. Of course, if $E\alpha a$ is a gradient extensor $F;\alpha a$, then a normalizing arc is also a stationary arc. (2) Then, in January 1968, Dr. Craig and G. H. Brigman published "An Extensor Generalization of a Multiple Integral Calculus-of-Variations Problem" in Volume 19 of Tensor.

In January 1970 I returned to Austin to complete my dissertation. Subsequent to its acceptance, I did some work on calculus of variations problems in which the base function $F(x, x', \ldots, x^{(M)})$ is a perfect derivative of some order of an underlying function f. I found that preferred linear combinations of extensors either vanished identically, or yielded Euler equations of appropriate order for the underlying function f. Then, in 1972, I co-authored a paper with Dr. Craig in Volume 25 of Tensor involving (1) primary extensors associated with a differentiated total energy function $[\,h\,(q, p)\,]^{(M)}$ and (2) the Hamiltonian equations of motion.

After I returned to NASA-Houston in 1971, I worked in the areas of post-flight analysis, trajectory optimization, advanced mission design, navigation, and pattern recognition analysis for the Large Area Crop Inventory Experiment. During the middle 1980s I was an adjunct instructor of mathematics at the University of Houston-Downtown. For a decade beginning in 1997, after my retirement from NASA, I was an adjunct instructor of mathematics at the University of Houston-Clear Lake. Meanwhile, my primary research interest turned to the automation of deep causal reasoning, to new accounts of backward causation, and to seeking possible manifestations of the new backward causation in quantum mechanics. In 1996 I published *Naïve Causal Modeling*, a book about these very topics. In 2009 I published, in Volume 22 of Physics Essays, the paper "Einstein-Podolsky-Rosen tachyons." In that paper I discussed how backward causation might operate, and how it might be mediated by superluminal particles in EPR experiments.

It appears that most research since 1950 involving extensors can be classified in one or more of three overlapping categories.

Category 1: **Applications of Extensors in Physics**

Much of the original work concerning applications of extensors in mathematical physics and the calculus of variations since 1950 seems to have been done by H. V. Craig. I have already cited most of that work, but will mention a later paper in Category 3.

Category 2: **Derivation of Extensors from Tensors and Other Extensors**

At times, mathematical operators have been used to derive new extensors from tensors or other extensors. A. Kawaguchi introduced certain operators in a cited 1940 paper.

P. S. Morey, Jr., in **18** (1967), generalized a Kawaguchi operator, then generalized again to include (1) multiple-parameter extensors, and (2) extensors under rheonomic transformations. He proceeded to derive new operators which produce extensors of essentially reduced range.

In **5** (1955), B. B. Townsend wrote about extensive derivatives of extensors. In **20** (1969), Robert H. Bowman and H. V. Craig discussed some properties of the extensive derivative. Robert H. Bowman, writing about differentiable extensions in two papers in **21** (1970), formulated a global setting for the concept of an extended tensor, as developed in the local case by H. V. Craig, A. Kawaguchi, and others. In **22** (1971), P. S. Morey, Jr. generalized to include weighted tensors certain formulas for construction of extensors by differentiation of tensors. In **28** (1974), Morey and J. Handy showed that certain quantities due to Zermelo are a special case of an operator which, when applied to an absolute excovariant extensor of sufficient differentiability, yields extensors of the same type, but of reduced range.

In **34** (1980), Craig obtained some extensors that are partly crossed and partly regular. He showed that contracting extensors with extensors derived from tensors by parameter differentiation yields an intrinsic derivative. Similarly, Craig showed how to obtain Lie type derivatives using basic Lie extensors. In **35** (1981), Morey found an excovariant analogue for a special Kawaguchi operator. Also in **35** (1981), Morey discussed horizontal contractions of second degree extensors. Morey wrote a second paper on horizontal contraction of second degree extensors in **36** (1982).

Category 3: **Generalization of the Extensor Concept**

Regarding generalization of the extensor concept, there were lines of research involving (a) multiple-parameter extensors, (b) g- extensors, (3) Jacobian extensors, (d) crossed extensors, (e) higher degree extensors, (f) spaces of non-integral order, (g) a global setting for the concept of extensors, and (h) complex extensors.

(a) **Multiple-parameter extensors**

Following A. Kawaguchi's introduction of multiple-parameter extensors in 1939 and Michiaki Kawaguchi's description in 1952 of multiple-parameter g-extensors and multiple-parameter extensors of separate order, H. V. Craig wrote about multiple-parameter Jacobian extensors in **2** (1952). I wrote about (1) multiple-parameter extensors with coupled-limit ranges and (2) certain applications in the calculus of variations in my doctoral dissertation (1971). H. V. Craig wrote about two-parameter extensors in the International Journal of Engineering Science **19** (1981). Toshiko Kawaguchi-Kanai and P. S. Morey, Jr. discussed crossed extensors in two parameters of separate order in **39** (1982), and T. Kawaguchi-Kanai dealt with properties of the extended point transformation in multiple parameters of separate order in Gakuen Review **26** (1975) and in the Journal of Hokkai-Gakuen **31** (1977).

(b) **G-extensors**

In a first paper in **2** (1952), Michiaki Kawaguchi introduced the g-extensor as a generalization of an extensor of A. Kawaguchi. In a second paper in the same volume, Michiaki Kawaguchi gave us multiple-parameter g-extensors, and multiple-parameter extensors of separate order. The same author proved some theorems applicable to g-extensors in **3** (1953), and examined covariant derivatives of g-extensors in **11** (1961). In **43** (1986) Mayuka F. Kawaguchi gave an alternate definition of g-extensor as a generalization of the extensor introduced by H. V. Craig in his 1937 paper. Mayuka F. Kawaguchi and Toshiko Kawaguchi-Kanai took a closer look at the relation between the two kinds of g-extensor in **56** (1995).

(c) **Jacobian Extensors**

H. V. Craig and W. T. Guy, Jr. introduced the notion of Jacobian extensors in the American Journal of Mathematics **72** (1950). As mentioned in (a), Craig described multiple-parameter Jacobian extensors in **2** (1952). Ch. Kano discussed crossed Jacobian extensors in **5** (1955). P. S. Morey, Jr. contributed higher degree Jacobian extensors in **30** (1976), in the process obtaining an intrinsic derivative of Jacobian extensors. Continuing to generalize the higher degree extensor concept, Morey offered higher degree Jacobian extensors in **33** (1979).

(d) **Crossed Extensors**

Extensors that are contravariant with respect to Greek indices and covariant with respect to Latin indices, or vice versa, are called crossed extensors. An example of a crossed extensor is the set of components of a covariant vector differentiated with respect to the curve parameter.

Crossed extensors were mentioned in H. V. Craig's 1943 textbook Vector and Tensor Analysis. H. Sasayama considered them in a 1953 paper. As we noted in (c), Ch. Kano defined crossed Jacobian extensors in **5** (1955). Yoshiharu Sato, in **32** (1978), found (1) mixed crossed extensors of arbitrary order and (2) their crossed extensors of reduced range. P. S. Morey, Jr., in **33** (1979), showed that crossed extensors may be regarded as extensors with negative indices. Morey and Craig found a one-to-one correspondence between regular extensors and crossed extensors in **36** (1982). . Mayuka F. Kawaguchi **43** (1986) concluded that g-extensor analysis is a generalization of ordinary crossed extensor analysis. Toshiko Kawaguchi-Kanai and Morey, in **49** (1990), applied Kawaguchi operators to crossed extensors. Toshiko Kawaguchi-Kanai, in **52** (1993), showed that a particular Kawaguchi operator can be applied to extensors in two parameters of separate order.

(e) **Higher degree extensors**

P. S. Morey, Jr. and H. V. Craig introduced second degree extensors, a generalization of the extensor concept, in **26** (1972). Morey, in **29** (1975) discussed higher degree extensors. Then, in **30** (1976) (as previously cited), Morey defined higher degree Jacobian extensors, and eventually obtained an intrinsic derivative of Jacobian extensors. In **33** (1979) Morey continued to investigate higher degree Jacobian extensors. In **35** (1981), and again in **36** (1982), Morey discussed horizontal contractions of second degree extensors. A horizontal contraction is a summation within a second degree extensor index which yields a first degree extensor or a tensor. Applicability of horizontal contraction has no analogue in standard extensor theory.

(f) **Spaces of non-integral order**

Hiroyoshi Sasayama wrote about extensors in a space of non-integral order in **3** (1953) and, in a separate paper in **3** (1953), about extensors of fractional grade.

(g) **A global setting for the concept of extensor**

Robert H. Bowman gave a global setting for the concept of an extended tensor in two previously cited (in Category 2) papers in **21** (1970).

(h) **Complex extensors**

In **26** (1972) H. V. Craig wrote about certain complex extensors whose components are either complex numbers or complex vectors. In **34** (1980) Craig found an application to relativity of extensors based on complex ordered pairs of Gibbs vectors.

CONCLUDING REMARKS

Based on my assessment of published research about extensors, I believe that my dissertation continues to offer contributions in all three of the stated categories of extensor research.

Chapter I, at least in its introduction via equation (12.8) of secondary extensors associated with a class c^{M+1} invariant function F $(x, x', \ldots, x^{(M)})$, impacts Category 2. Furthermore, Chapter's I's description of multiple-parameter extensors with coupled-limit ranges extends Category 3.

Chapters II and VI significantly expand the body of research in Category 1 concerning extensors and the calculus of variations. Chapter II, by introducing three-parameter extensors based upon a variation parameter and two parameters from an auxiliary plane, affects the line of research in Category 3 involving multiple-parameter extensors. By applying these three-parameter extensors to the higher variations in a simple double integral problem of the calculus of variations, Chapter II also enlarges Category 1.

Chapter III, by describing use of differentiation to construct an excovariant extensor associated with the function $f(x, x', \ldots, x^{(M)}, y\Theta a)$ – with f invariant of weight zero and y an absolute excovariant extensor – contributes to Category 2. Chapter III then goes on to extend Category 1 by finding an application of the constructed extensor in Hamiltonian dynamics.

Chapters IV, V, and VI impact Category 2 by investigating linearity properties within a family of extensors derived from the non-tensor ranks of an arbitrary excovariant extensor $E\alpha a$ for both (1) a single-parameter case α: 0 to M, and (2) a two-parameter case with (α_1, α_2) having range $\alpha_1 \geq 0, \alpha_2 \geq 0, \alpha_1 + \alpha_2 \leq M$.

Therefore, to enhance the accessibility of the fresh material in *On Extensor Structures in the Calculus of Variations*, that document is now presented in book form, and by its publishing, provides it the broader distribution that it heretofore lacked.

William Clifton Bean
August, 2015

PREFACE

It has been found that a single linear extensor equation expresses the Euler differential equations which constitute a necessary and sufficient condition that the arc C_0 is a stationary arc in calculus of variations problems of the type $\int_{t_1}^{t_2} F(x,x', \ldots ,x^{(M)})dt$. Here x is the set of N coordinate variables $x^1, x^2, \ldots, x^N$, and parenthetical superscripts and primes (') indicate differentiation with respect to the curve parameter t.

It has further been observed that the vanishing of the Hamiltonian tensor associated with the extensor $E_{\alpha a}$ is a necessary and sufficient condition that the arc C_0 is a normalizing arc in a modified calculus of variations problem in which there is investigated _ab initio_ the requirement $\int_{t_1}^{t_2} \sum_{\alpha=0}^{M} E_{\alpha a}V^{a(\alpha)}dt = 0$, rather than the vanishing of the first variation of an integral. Here $E_{\alpha a}$ is an arbitrary excovariant extensor of range α: 0 to M, and V^a is a variation tensor which generates a family of admissible comparison arcs. Also, analogous results have been obtained for a certain extensor-generalized double integral calculus of variations problem.

Because the calculus of variations problems to be considered involve integrals dependent upon such geometric entities as parameterized arcs and surfaces, and because tensors and extensors can be defined along parameterized arcs or surfaces it is natural to anticipate a further role for extensors in the calculus of variations. A first question arises: (1) Is it possible to extend to higher variations the result that the first variation of $\int_{t_1}^{t_2} F(x, x', \ldots, x^{(M)})dt$ can be expressed as the integral of an extensor contraction? A second question occurs: (2) Is there an extensor formulation applicable to Hamiltonian dynamics based upon the function $F(x, x', \ldots, x^{(M)}, y_{\theta a})$, y an excovariant extensor, as there is upon the function $f(x, y_a)$, y a covariant tensor? A further question appears: (3) Are there tensors other than the Hamiltonian tensor which may be associated with an extensor or matrix-extensor $E_{\alpha a}$? If so, two additional questions arise with regard to the interpretation of these tensors: (4) What is the relationship of the set of tensors associated with $E_{\alpha a}$ to the variational integral $\int_{t_1}^{t_2} \sum_{\alpha} E_{\alpha a} V^{a(\alpha)} dt$, or $\iint_{D_0} \sum_{\alpha} E_{\alpha a} V^{a(\alpha)} du^1 du^2$? (5) How may the set of tensors associated with $E_{\alpha a}$ be applied to a class of variational integrals which may be regarded as generated by $E_{\alpha a}$? The purpose of this dissertation is to investigate these questions.

The author wishes to express gratitude to Professor H. V. Craig, whose continued encouragement, advice, patience, and assistance have made possible the development of this work. Also, he would like to thank Mr. David Nilsson for his helpful suggestions in the revision of this manuscript. He is indebted to Professors E. J. Prouse, W. T. Guy, Jr., and E. L. Hudspeth for their review of the manuscript in completed form.

William Clifton Bean

ON CERTAIN EXTENSOR STRUCTURES

IN THE CALCULUS OF VARIATIONS

ABSTRACT

The following topics regarding extensor structures in the calculus of variations are developed: (1) Two different types of summation ranges applicable to two-parameter extensors are distinguished, and the associated imbedding theorems are examined. (2) The role of a set of generalized Synge tensors relative to the invariance under parameter transformations of certain single and double integrals of the calculus of variations is studied. The generalized Synge tensors are obtained by replacing the gradient extensor with a general extensor $E_{\alpha a}$. (3) Extensor structures are presented for the higher variations of the integral $\int_{t_1}^{t_2} F(x, x', \ldots, x^{(M)})dt$ of the ordinary simple calculus of variations problem of the indicated type. A two-parameter extensor-based Kawaguchi theorem is first proved, then applied to the expression of the higher variations of $\int_{t_1}^{t_2} F(x, x', \ldots, x^{(M)})dt$ as integrals of contractions of two-parameter extensors. Here, the two parameters are the curve parameter t and the variation parameter v, respectively. A recursion result is then noted, namely that the integrand of the (K+1)st variation, $K \geq 1$, is precisely a contraction of a two-parameter variation extensor with the two-parameter gradient extensor associated with the integrand of the Kth variation. Analogous results involving three-parameter extensors for the simple double integral $\iint_{D_0} F(x, x^{(0,1)}, x^{(1,0)}, \ldots, x^{(0,M)}, \ldots, x^{(M,0)})du^1du^2$ are given, and a multiple-parameter construction for higher derivatives of the basic integral in extensor-generalized problems of the calculus of variations is noted.

(4) There is introduced a generalization involving the function $f(x, x', \ldots, x^{(M)}, y_{\theta a})$ with f invariant of weight zero under coordinate transformations and y excovariant. The basic presumption of the equivalence of a certain pair of excontravariant extensors, namely $\frac{\partial f}{\partial y_{\alpha a}} = x'^{(\alpha)a}$ for α : 0 to M, along the arc in question is shown to lead to the conclusion that $F_{\alpha a}$ is an extensor of range α : 0 to M + 1, for $F_{\alpha a} = -\left.\frac{\partial f}{\partial x^{(\alpha)a}}\right|_y + y_{(\alpha-1)a}$, $\alpha = 1, 2, \ldots, M$, $F_{0a} = -\frac{\partial f}{\partial x^a}$, $F_{(M+1)a} = y_{Ma}$. The Kawaguchi extensor $S_{\alpha a}$, of range α : 0 to M + 1,

defined by $S_{\alpha a} = \binom{M+1}{A} y_{Ma}{}^{\cdot(M+1-A)}$, $A = \alpha$, for $\alpha = 0, 1, \ldots, M + 1$ is introduced. The quantities $F_{\alpha a}$ and $S_{\alpha a}$ are called the primary extensors associated with $f(x, x', \ldots, x^{(M)}, y_{\theta a})$.

(5) A generalized Legendre transformation is introduced in which the Legendre transform $F(x, x', \ldots, x^{(M)}, x^{(M+1)})$ of $f(x, x', \ldots, x^{(M)}, y_{\theta a})$ is defined by $F \equiv \sum_{\alpha=0}^{M} x'^{(\alpha)a} y_{\alpha a} - f$. The primary extensors $F_{\alpha a}$ and $S_{\alpha a}$ associated with f are seen to be equal, respectively, to the primary extensors $F_{;\alpha a}$ and $R_{\alpha a}\left(\equiv \binom{M+1}{A} F_{;(M+1)a}{}^{\cdot(M+1-A)}{}_{,A=\alpha}\right)$ associated with F.

(6) A possible generalization of the concept of primary extensor applied to a function of the type $f(x, x', \ldots, x^{(M)}, y_{\theta a})$ is treated briefly, wherein it is shown that for each integral value of ϕ, the quantities $\phi_{\alpha a}$ (and the associated Kawaguchi extensors which could be constructed therefrom) could be regarded as generalized primary extensors (of range α : 0 to $M + 1 - \phi$) associated with f. The quantities $\phi_{\alpha a}$ are defined by

$$\phi_{\alpha a} = -\binom{\alpha + \phi}{\phi} \frac{\partial f}{\partial x^{(\alpha+\phi)a}} + \binom{\alpha + \phi - 1}{\phi} y_{(\alpha+\phi-1)a}$$

for $\alpha = 1, 2, \ldots, M - \phi$, $\phi_{0a} = -\frac{\partial f}{\partial x^{(\phi)a}}$, $\phi_{(M+1-\phi)a} = y_{Ma}$.

(7) A dynamical application occurs when $h^{(M)}$, the Mth derivative (with $M > 0$) of the total energy function $h(q,p)$ in Hamiltonian form, is adopted as a basic function. As a first observation, we see that the Kawaguchi extensor $K_{\alpha a}\left(\equiv \binom{M}{A} p_a^{\cdot (M-A)},\ A = \alpha\right)$ of range α : 0 to M based on the momentum tensor p_a is identical to the gradient extensor $P_{\alpha a}\left(\equiv \frac{\partial}{\partial q^{,(\alpha)a}} T^{(M)}\right)$ of range α : 0 to M. We see further that the Hamiltonian equations $p_a' = -\frac{\partial h}{\partial q^a}$ (and parameter derivatives of these equations) are expressible as the equality of the primary extensors $E_{\alpha a}$ and $W_{\alpha a}$ associated with $h^{(M)}$. Here, the primary extensors $E_{\alpha a}$ and $W_{\alpha a}$ are of range α : 0 to M + 1 and are defined in accordance with the fact that $h^{(M)}$ is a function of $q, q', \ldots, q^{(M)}$, and the excovariant quantities $P_{\alpha a}$. The basic presumption $\frac{\partial h^{(M)}}{\partial P_{\alpha a}} = q^{,(\alpha)a}$ for α : 0 to M is identically satisfied along all trajectories in our dynamical space.

We then have $E_{\alpha a} = -\frac{\partial h^{(M)}}{\partial q^{(\alpha)a}} + P_{(\alpha-1)a}$ for $\alpha = 1, 2, \ldots,$ M, $E_{0a} = -\frac{\partial h^{(M)}}{\partial q^{a}}$, $E_{(M+1)a} = P_{Ma} = p_a$. We also have $W_{\alpha a} = \binom{M+1}{A} p_a^{\cdot(M+1-A)}$, $A = \alpha$, for $\alpha = 0, 1, \ldots, M+1$. The Hamiltonian equations of motion are also seen to be obtainable as the Euler (or Hamiltonian) equations associated with the simple extensor-generalized calculus of variations problem based upon $\int_{t_1}^{t_2} \sum_{\alpha=0}^{1} A_{\alpha a} V^{a(\alpha)} dt$, where $A_{\alpha a} : -\left.\frac{\partial h}{\partial q^{a}}\right|_{p}$, p_a is the first primary extensor associated with $h(q,p)$. The Euler equations associated with the simple extensor-generalized calculus of variations problem based upon $\int_{t_1}^{t_2} \sum_{\alpha=0}^{M+1} E_{\alpha a} V^{a(\alpha)} dt$ vanish identically.

(8) A further topic considered is the problem, given a one-parameter or two-parameter extensor $E_{\alpha a}$, to construct therefrom a family $\mathscr{S}(E_{\alpha a})$ of "fundamental extensors associated with $E_{\alpha a}$." Such a family is generated. The family $\mathscr{S}(E_{\alpha a})$ is partitioned into subsets, and certain associated linearity properties are stated. A set of Kawaguchi base tensors associated with $\mathscr{S}(E_{\alpha a})$ is obtained, as is a set of Hamiltonian-type tensors. These sets are shown to be equivalent to each other and to a set of Synge tensors associated with $E_{\alpha a}$. Now

classes of extensor-generalized single and double integral calculus of variations problems are introduced, with respect to which the Synge tensors are found to describe normalizing arcs.

TABLE OF CONTENTS

C H A P T E R   I

INTRODUCTION TO EXTENSORS AND RELATED TOPICS

1. Notation, nomenclature and the summation convention. The notation used in this dissertation is basically the conventional notation of tensor and extensor analysis, as in [5] and [6].* In addition the notation of multiple parameter theory is like that of [13] and [14], except that a two-parameter theory involving coupled limits is developed, and certain comparisons made with the uncoupled limit theory from the standpoint of imbedding. Further, certain basic notation of elementary set theory is adopted. In general barred and unbarred symbols are associated with the coordinate systems $\bar{x}$ and x, respectively. The Latin and Greek indices $a, b, c, \ldots \alpha, \beta, \gamma, \ldots$ refer to the x-system, whereas $r, s, t, \ldots \rho, \sigma, \tau, \ldots$ refer to the $\bar{x}$-system. Because the coordinate systems ordinarily may be distinguished by the letters used as indices, the bar may be omitted.

---

*Numbers in brackets refer to items in the appended bibliography.

Differentiation with respect to the parameter t of a parameterized arc is designated by primes and enclosed Greek or numerical indices, and partial derivatives are represented by means of subscripts as follows (see [5], p. 210):

$$
(1.1)\quad \begin{cases}
x^{(0)a} = x^a \;;\quad x'^a = x^{(1)a} = \dfrac{dx^a}{dt} \;;\quad x^{(\alpha)a} = \dfrac{d^\alpha x^a}{dt^\alpha} \;; \\[2ex]
X^r_a = X^{(0)r}_{(0)a} = \dfrac{\partial x^r}{\partial x^a} \;;\quad F_{;\alpha a} = \dfrac{\partial F}{\partial x^{(\alpha)a}} \;;\quad X^{\rho r}_{\alpha a} = \dfrac{\partial x^{(\rho)r}}{\partial x^{(\alpha)a}} \;; \\[2ex]
X^{r(\rho)}_a = \dfrac{d^\rho X^r_a}{dt^\rho} \;;\quad F^{(\rho)}_{;\alpha a} = \dfrac{\partial}{\partial x^{(\alpha)a}}\left(\dfrac{d^\rho F}{dt^\rho}\right) \;; \\[2ex]
F_{;\alpha a}{}^{(\rho)} = \dfrac{d^\rho}{dt^\rho}\left(\dfrac{\partial F}{\partial x^{(\alpha)a}}\right) .
\end{cases}
$$

The symbol $\binom{A}{P}$ is defined by $\binom{A}{P} = \frac{A!}{P!(A-P)!}$, subject to the conventions $\binom{A}{0} = 1$, $\binom{0}{0} = 1$, and $\binom{A}{P} = 0$ for $A < P$. The Kronecker delta symbol δ^A_B is defined by: $\delta^A_B = 1$ for $A = B$; $\delta^A_B = 0$ for $A \neq B$. It should be explicitly noted that, unless stated otherwise, symbols with out-of-range indices will be assigned the value zero.

Sufficient generality is obtained for the case of multiple parameters by taking a set of two parameters u^1, u^2. Partial differentiation with respect to the parameters u^1, u^2 is designated by one-row matrix primes. Thus we adopt the definitions:

$$
(1.2)\quad
\begin{cases}
\alpha^+ = \alpha_1 + \alpha_2 \quad ; \\[4pt]
(\alpha) = (\alpha_1, \alpha_2) \quad ; \\[4pt]
\begin{pmatrix}\alpha\\ \beta\end{pmatrix} = \begin{pmatrix}\alpha_1\\ \beta_1\end{pmatrix}\begin{pmatrix}\alpha_2\\ \beta_2\end{pmatrix}, \quad \begin{pmatrix}\alpha\\ \beta\end{pmatrix} = 0 \text{ if } \\
\qquad\qquad \alpha_1 < \beta_1 \text{ or } \alpha_2 < \beta_2 ; \\[4pt]
\sum_{\alpha_1+\alpha_2 \geq 0}^{\alpha_1+\alpha_2 \leq M} \ : \ \alpha \text{ runs over the set of matrices } (\alpha_1, \alpha_2) \\
\qquad\qquad \text{such that } \alpha_1 \geq 0,\ \alpha_2 \geq 0,\ \alpha_1 + \alpha_2 \leq M ; \\[4pt]
(-1)^\alpha = (-1)^{\alpha_1} (-1)^{\alpha_2} \quad ; \\[4pt]
\delta^\alpha_\beta = \delta^{\alpha_1}_{\beta_1} \delta^{\alpha_2}_{\beta_2} \quad ; \\[4pt]
F^{(\alpha_1, \alpha_2)} = \dfrac{\partial^{\alpha_1+\alpha_2} F}{\partial (u^1)^{\alpha_1} \partial (u^2)^{\alpha_2}} \\
\qquad\qquad \text{for } F \text{ of class } C^{\alpha^+} .
\end{cases}
$$

Also useful is the property involving matrix addition:

$$
(1.3)\quad (F^{(\alpha)})^{(\beta)} = F^{(\alpha+\beta)} = F^{(\alpha_1+\beta_1, \alpha_2+\beta_2)} .
$$

The definitions of symbols given by equations (1.1) may be interpreted for the matrix case without formal change.

With regard to set theory, we regard as primitive the notion of a set as a collection of things. The notation $a \in S$ is read "a is an element of the set S." The notation $A = \{a \in S | P(a)\}$ is read "A is the set of all elements in S for which the property P holds." If every element of the set A is also an element of the set S, then the set A is said to be a subset of the set S, written $A \subset S$. The sets A and B are said to be equal, written $A = B$, if both $A \subset B$ and $B \subset A$. The union of the two sets A and B, written $A \cup B$, is the set $\{x | x \in A \text{ or } x \in B\}$. The intersection of the two sets A and B, written $A \cap B$ is the set $\{x | x \in A \text{ and } x \in B\}$. The difference set $A - B$ is the set $\{x \in A | x \notin B\}$. A set S is said to be countably infinite if there exists a one-to-one mapping from S to the set of positive integers. The concepts of set union and set intersection can be extended to arbitrary finite or countably infinite collections of sets.

With regard to nomenclature, the word 'vector' is used herein in more than one sense. To emphasize that a certain object has the mathematical properties of a directed stroke we will refer to it as a 'Gibbs vector.' An entity which is 'an infinite collection of sets of N-labeled numbers which follow either the contravariant or the covariant laws under admissible coordinate transformations' will be called simply a 'vector.' An 'element of a set having the combinatorial and existence properties of an algebraic linear vector space' will be called a 'Vector.' Henceforth, capitalization will also be used to designate such a space as an 'algebraic linear Vector space.' The various usages of the term 'vector' are not mutually exclusive, e.g., a Gibbs vector could be both a vector and a Vector.

A summation convention amended from [5], p. 210, is used: (a) repeated lower-case Latin indices generate summations from 1 to N; (b) repeated lower-case Greek indices do not generate summations, unless the contrary is indicated by the presence of a summation sign giving the range; (c) Greek indices may occasionally be replaced by capitals, as ρ by P, in order to emphasize that summation is not intended. In the case of two parameters repeated lower-case Greek indices do not generate summations.

2. Leibnitz rule. The Leibnitz rule for differentiation of two class c^M functions U and V with respect to a single parameter t may be written as follows:

$$(2.1)\qquad (UV)^{(M)} = \sum_{\alpha=0}^{M} \binom{M}{\alpha} U^{(\alpha)}V^{(M-\alpha)} \quad .$$

The formula (2.1) is valid for the case of multiple parameters, provided of course that 0, α, and M denote the appropriate matrices; specifically, we have the following:

Theorem (2.1). If $U(u^1,u^2)$, $V(u^1,u^2)$ are each of class c^{M^+}, $M^+ = M_1 + M_2$, and if 0, α, and M denote the matrices (0,0), (α_1,α_2), and (M_1,M_2), respectively, then

$$(UV)^{(M)} = \sum_{\alpha_1+\alpha_2\geq 0}^{\alpha_1+\alpha_2\leq M^+} \binom{M}{\alpha} U^{(\alpha)}V^{(M-\alpha)} \quad .$$

Proof: Differentiating partially two class c^{M^+} functions U and V with respect to the parameters u^1 and u^2, we have by formula (2.1) and the definitions (1.2) that

$$(UV)^{(M)} = (UV)^{(M_1, M_2)}$$

$$= \left[(UV)^{(M_1,0)}\right]^{(0,M_2)}$$

$$= \left[\sum_{\alpha_1=0}^{M_1} \binom{M_1}{\alpha_1} U^{(\alpha_1,0)} V^{(M_1-\alpha_1,0)}\right]^{(0,M_2)}$$

$$= \sum_{\alpha_1=0}^{M_1} \binom{M_1}{\alpha_1} \sum_{\alpha_2=0}^{M_2} \binom{M_2}{\alpha_2} \left[U^{(\alpha_1,0)}\right]^{(0,\alpha_2)} \left[V^{(M_1-\alpha_1,0)}\right]^{(0,M_2-\alpha_2)}$$

$$= \sum_{\alpha_1=0}^{M_1} \sum_{\alpha_2=0}^{M_2} \binom{M_1}{\alpha_1} \binom{M_2}{\alpha_2} U^{(\alpha_1,\alpha_2)} V^{(M_1-\alpha_1, M_2-\alpha_2)}$$

$$= \sum_{\alpha_1+\alpha_2\geq 0}^{\alpha_1+\alpha_2\leq M^+} \binom{M}{\alpha} U^{(\alpha)} V^{(M-\alpha)}$$

where we note that terms for which $\alpha_1 > M_1$ or $\alpha_2 > M_2$, and yet $\alpha_1 + \alpha_2 \leq M_1 + M_2$, vanish by virtue of the presence of the binomial coefficient $\binom{M}{\alpha}$.

3. <u>The extended coordinate transformation</u>. (See [6], pp. 213, 214). Definition (3.1). Let

$$(3.1) \qquad \begin{cases} x^a = x^a(x^1, x^2, \ldots, x^N) = x^a(\bar{x}) \\ \bar{x}^r = \bar{x}^r(x^1, x^2, \ldots, x^N) = \bar{x}^r(x) \end{cases}$$

be a complete coordinate transformation which maps the domains D and $\bar{D}$ of auxiliary coordinate spaces one-to-one reversibly. Further, let the functions $x^a(\bar{x})$, $\bar{x}^r(x)$ be of class c^M in D and $\bar{D}$, respectively. Finally, let C be a parameterized arc $x^a = x^a(t)$ which for t in an interval (t_1,t_2) is of class c^M and lies in D. The transformation equations obtained for $x^{(\lambda+1)}$, $\lambda = 0, 1, \ldots, M-1$, by repeatedly differentiating (3.1) with respect to the parameter t, together with the original equations (3.1) constitute an extended coordinate transformation of order M, which may be written

$$
(3.2)\quad \left\{
\begin{array}{lll}
\bar{x}^r & = \bar{x}^r(x) & , \\
\bar{x}^{(\rho+1)r} & = \sum_{\alpha=0}^{\rho} \binom{\rho}{\alpha} X_a^{r(\rho-\alpha)} x^{(\alpha+1)a} & , \quad \rho = 0,1,\ldots,M-1; \\
x^a & = x^a(\bar{x}) & , \\
x^{(\alpha+1)a} & = \sum_{\rho=0}^{\alpha} \binom{\alpha}{\rho} X_r^{a(\alpha-\rho)} \bar{x}^{(\rho+1)r} & , \quad \alpha = 0,1,\ldots,M-1.
\end{array}
\right.
$$

These equations may be applied to the multiple parameter extended coordinate transformation without formal change.

Definition (3.2). The transformation equations obtained for $x^{(0,1)}, x^{(1,0)}, \ldots, x^{(0,M)}, \ldots, x^{(M,0)}$ by repeated partial differentiation of equations (3.1) where now each x^a is a class c^M function of the two parameters u^1, u^2, together with the original equations (3.1) constitute a two-parameter extended coordinate transformation. That the resulting equations are formally the same as those of (3.2) follows from the fact that the form of the Leibnitz rule is retained for the two-parameter case, as established in Section 2.

4. Certain properties of $X^{\rho r}_{\alpha a}$. A very important property of the symbols $X^{\rho r}_{\alpha a}$ is expressed by the following reduction formula:

(4.1) $$X^{\rho r}_{\alpha a} = \binom{\rho}{\alpha} X^{r(\rho-\alpha)}_{a} \quad \text{where } \alpha \leq \rho \quad .$$

The validity of this relationship may be established by an induction, as in [5], p. 215, the steps of which apply to the matrix form. Similarly, in case $\alpha > \rho$ we have the relationship

(4.2) $$X^{\rho r}_{\alpha a} = 0 \qquad \text{if } \rho < \alpha \quad .$$

Formulas (4.1) and (4.2) carry over to the case of matrix primes (see [13], p. 30 and [14], p. 28) without formal change except that the requirement that $\rho < \alpha$ in (4.2) should be replaced by the stipulation ρ is 'partly less' than α. The statement that ρ is 'partly less' than α means that there exists an i such that $\rho_i < \alpha_i$. In the two-parameter case i is, of course, restricted to the set 1, 2, and formulas (4.1) and (4.2) become, respectively,

$$X^{(\rho_1,\rho_2)r}_{(\alpha_1,\alpha_2)a} = \binom{\rho_1}{\alpha_1}\binom{\rho_2}{\alpha_2} X^{r(\rho_1-\alpha_1,\rho_2-\alpha_2)}_a ,$$

$$X^{(\rho_1,\rho_2)r}_{(\alpha_1,\alpha_2)a} = 0 \quad \text{if } \rho_1 < \alpha_1 \text{ or } \rho_2 < \alpha_2 .$$

The following important formulas are also well known:

$$(4.3) \qquad X^{(Q)r}_{(Q)a} = X^r_a ,$$

$$(4.4) \qquad X^{\rho r}_{\alpha a} X^{\alpha a}_{\sigma s} = \delta^{\rho}_{\sigma} \delta^r_s .$$

Since by definition $\delta^{\beta}_{\gamma} = \delta^{\beta_1}_{\gamma_1} \delta^{\beta_2}_{\gamma_2}$, where $\beta = (\beta_1,\beta_2)$ and $\gamma = (\gamma_1,\gamma_2)$, both (4.3) and (4.4) extend to the case of matrix primes.

5. The extensor transformation law; extensor ranks.

Definition (5.1). Components of an extensor, a typical case (see [5], pp. 260-262): If there is given at a point P of a parameterized arc of class c^M one set of quantities $E^{\alpha a \cdot b}_{\gamma c \cdot d}$ for each coordinate system x of our infinite collection of (3.1), and if these labeled quantities (which may be either numbers or Gibbs vectors) are such that for every pair $x, \bar{x}$ of coordinate systems the associated quantities $E^{\alpha a \cdot b}_{\gamma c \cdot d}$, $\bar{E}^{\rho r \cdot s}_{\tau t \cdot u}$ are related by the transformation law

$$(5.1) \qquad E^{\alpha a \cdot b}_{\gamma c \cdot d} = J^W\left(\frac{\bar{x}}{x}\right) \bar{E}^{\rho r \cdot s}_{\tau t \cdot u}\, X^{\alpha a}_{\rho r} X^{b}_{s} X^{\tau t}_{\gamma c} X^{u}_{d} ,$$

then these labeled quantities are the components of an extensor of range M, weight W, excontravariant order one, excovariant order one, contravariant order one, and covariant order one. If the quantities E instead of being associated with P are functions of the sets of variables x , . . . , $x^{(Q)}$, then the labeled quantities are said to be the components of an extensor of functional order Q and of characteristic (1, 1, 1, 1, W, M, Q).

The definition of multiple parameter extensor may be obtained from definition (5.1) by letting the Greek indices denote the appropriate matrices with no formal change in the transformation equation (5.1) (see [13]). For brevity the term 'extensor' often will be used with reference to the symbol for its components. An extensor for which $W = 0$ is said to be an absolute extensor.

It is noteworthy in the special case in which only doublet indices are involved that extensor components may be interpreted as tensor components in a space of $N(M+1)$ dimensions for the one-parameter case or of $\frac{N(M+1)(M+2)}{2}$ dimensions for the two-parameter case, so that the simplest concept of extensor is that of tensor relative to the extended coordinate transformation (see [5]).

The fact that the extended coordinate transformation has a polynomial structure in the primed variables ensures the existence of special properties of extensors for which no tensor analogs exist. By virtue of the property of the symbols $X^{\alpha a}_{\rho r}$ and $X^{\tau t}_{\gamma c}$ given by equation (4.2), the x-components of E for which $\alpha \leq \theta$ are independent of the

$\bar{x}$-components having $\rho > \theta$, and the x-components for which $\gamma \geq \phi$ depend only on those $\bar{x}$-components for which $\tau \geq \phi$. Consideration of extensor components which meet such restrictions leads to the concepts 'ranks' and 'extensor components of reduced range.' For a discussion of these concepts for the one-parameter case, see [5] or [6], pp. 268-275. Matrix extensors of reduced range occur in the same way, where the requirements $\alpha \leq \theta$ and $\rho > \theta$ are replaced by obvious modifications.

The components of extensors may be grouped naturally into ranks of N members each according to the values of the Greek indices. Thus $E_{\alpha a}$ has the M+1 ranks $E_{0a}, E_{1a}, \ldots, E_{Ma}$, or $E_{(\alpha_1,\alpha_2)a}$, $\alpha_1 + \alpha_2 \leq M$, has the $\frac{(M+1)(M+2)}{2}$ ranks $E_{(0,0)a}, \ldots, E_{(0,M)a}, \ldots, E_{(M,0)a}$.

Theorem (5.1). If $E_{\alpha a}$ is an excovariant extensor of range α : 0 to M in the one-parameter case, then E_{Ma} is a tensor rank. If $E_{\alpha a}$ is a two-parameter excovariant extensor of range $\alpha_1 \geq 0$, $\alpha_2 \geq 0$, $\alpha_1 + \alpha_2 \leq M$, then all $E_{(\alpha_1, M-\alpha_1)a}$ such that $0 \leq \alpha_1 \leq M$ are tensor ranks.

Proof: By the extensor transformation law we have

$$E_{Ma} = \sum_{\rho=0}^{M} E_{\rho r} X^{\rho r}_{Ma}$$

in the one-parameter case. By property (4.2) $X^{\rho r}_{Ma} = 0$ for $\rho < M$, implying

(5.2) $$E_{Ma} = E_{Ma} X^{Mr}_{Ma} = E_{Ma} X^{r}_{a} \quad ,$$

i.e., E_{Ma} transforms as a tensor.

In the two-parameter case, again by the extensor transformation law, we may write

(5.3) $$E_{(\alpha_1, M-\alpha_1)a} = \sum_{\rho_1+\rho_2 \geq 0}^{\rho_1+\rho_2 \leq M} E_{\rho r} X^{\rho r}_{(\alpha_1, M-\alpha_1)a} \quad .$$

Now the index ρ of summation runs over matrices (ρ_1, ρ_2) for which $\rho_1 + \rho_2 \leq M$. If $\rho_1 + \rho_2 < M$, then it is not possible that both $\rho_1 \geq \alpha_1$ and $\rho_2 \geq M - \alpha_1$. Therefore, for $\rho_1 + \rho_2 < M$ by property (4.2) $X^{\rho r}_{(\alpha_1, M-\alpha_1)a}$ vanishes. If $\rho_1 + \rho_2 = M$ we may satisfy both $\rho_1 \geq \alpha_1$ and $\rho_2 \geq M - \alpha_1$ only if $\rho_1 = \alpha_1$ and $\rho_2 = M - \alpha_1$. Therefore, the only non-vanishing term in the right member of (5.3) occurs for $\rho_1 = \alpha_1$, $\rho_2 = M - \alpha_1$, and we have

$$(5.4) \qquad E_{(\alpha_1,M-\alpha_1)a} = E_{(\alpha_1,M-\alpha_1)r}X^{(\alpha_1,M-\alpha_1)r}_{(\alpha_1,M-\alpha_1)a}$$

$$= E_{(\alpha_1,M-\alpha_1)r}X^r_a \quad ,$$

i.e., $E_{(\alpha_1,M-\alpha_1)a}$, $\alpha_1 = 0, 1, \ldots M$, transforms as a tensor.

Theorem (5.2). If the tensor rank E_{Ma} of the one-parameter excovariant extensor $E_{\alpha a}$ of range α : 0 to M vanishes identically, the rank $E_{(M-1)a}$ next removed becomes a tensor rank. If the tensor ranks $E_{(\alpha_1,M-\alpha_1)a}$, $\alpha_1 = 0, 1, \ldots, M$, of the two-parameter excovariant extensor $E_{(\alpha_1,\alpha_2)a}$ of range $\alpha_1 \geq 0$, $\alpha_2 \geq 0$, $\alpha_1 + \alpha_2 \leq M$, all vanish identically, then the ranks $E_{(\alpha_1,M-1-\alpha_1)a}$, $\alpha_1 = 0, 1, \ldots, M - 1$, next removed become tensor ranks.

Proof (one-parameter case):

$$E_{(M-1)a} = \sum_{\rho=0}^{M} E_{\rho r}X^{\rho r}_{(M-1)a}$$

$$= E_{(M-1)r}X^{(M-1)r}_{(M-1)a} + E_{Mr}(\equiv 0)X^{Mr}_{(M-1)a}$$

$$= E_{(M-1)r}X^r_a \quad .$$

Proof (two-parameter case):

$$E_{(\alpha_1,M-1-\alpha_1)a} = \sum_{\rho_1+\rho_2\geq 0}^{\rho_1+\rho_2\leq M} E_{(\rho_1,\rho_2)r}X^{(\rho_1,\rho_2)r}_{(\alpha_1,M-1-\alpha_1)a} .$$

Here the index of summation ρ runs over matrices (ρ_1,ρ_2) for which $\rho_1 + \rho_2 \leq M$. If $\rho_1 + \rho_2 < M - 1$, then it is not possible that both $\rho_1 \geq \alpha_1$ and $\rho_2 \geq M - 1 - \alpha_1$. Therefore, for $\rho_1 + \rho_2 < M - 1$, by property (5.2), $X^{(\rho_1,\rho_2)r}_{(\alpha_1,M-1-\alpha_1)a}$ vanishes. Furthermore, if $\rho_1 + \rho_2 = M-1$ we may satisfy both $\rho_1 \geq \alpha_1$ and $\rho_2 \geq M - 1 - \alpha_1$ only if $\rho_1 = \alpha_1$ and $\rho_2 = M - 1 - \alpha_1$. We may now write

$$E_{(\alpha_1,M-1-\alpha_1)a} = E_{(\alpha_1,M-1-\alpha_1)r}X^{(\alpha_1,M-1-\alpha_1)r}_{(\alpha_1,M-1-\alpha_1)a}$$

$$+ \sum_{\beta=0}^{M} E_{(\beta,M-\beta)r}(\equiv 0)X^{(\beta,M-\beta)r}_{(\alpha_1,M-1-\alpha_1)a}$$

$$= E_{(\alpha_1,M-1-\alpha_1)r}X^{r}_{a} .$$

By induction, if ranks M, M-1, . . . , A+1 vanish in the one-parameter case, then rank A is a tensor rank. Similarly, if all ranks (α_1,α_2) such that $\alpha_1 + \alpha_2 \geq A + 1$ vanish in the two-parameter case, then all ranks (α_1,α_2) such that $\alpha_1 + \alpha_2 = A$ are tensor ranks.

6. Extensor character of $F_{;\alpha a}$ and $\binom{M}{A} T_a{}^{\cdot(M-A)}$.

Theorem (6.1). If $F(x, \ldots, x^{(M)})$ is a class c' invariant function, i.e., if

$$(6.1) \qquad \bar{F}(\bar{x}, \ldots, \bar{x}^{(M)}) \equiv F\left[x(\bar{x}), \ldots, x^{(M)}(\bar{x}, \ldots, \bar{x}^{(M)})\right]$$

is of class c', then $F_{;\alpha a}$ are the components of an extensor.

Proof: Differentiating (6.1) with respect to the x's, we obtain

$$(6.2) \qquad \bar{F}_{;\rho r} = \sum_{\alpha=0}^{M} F_{;\alpha a} X^{\alpha a}_{\rho r}$$

so that the quantities $F_{;\alpha a}$ follow the extensor transformation law of excovariant order unity. The method of proof can be applied to the multiple-parameter case.

Theorem (6.2). Kawaguchi theorem. If the quantities T_a are the components of an absolute covariant tensor of class c^M (and of class c^{M^+}, $M^+ = M_1 + M_2$), then we have both (1) the quantities $\binom{M}{\alpha} T_a{}^{\cdot(M-\alpha)}$ are the components of a one-parameter excovariant extensor of range α : 0 to M, and (2) the quantities $\binom{M_1}{\alpha_1}\binom{M_2}{\alpha_2} T_a{}^{\cdot(M_1-\alpha_1, M_2-\alpha_2)}$

are the components of a two-parameter excovariant extensor of range $0 \leq \alpha_1 \leq M_1$, $0 \leq \alpha_2 \leq M_2$ (which can be imbedded in the range $\alpha_1 \geq 0$, $\alpha_2 \geq 0$, $\alpha_1 + \alpha_2 \leq M^+$ by assigning the values zero to those ranks (α_1, α_2) satisfying $\alpha_1 + \alpha_2 \leq M^+$ for which either $\alpha_1 > M_1$ or $\alpha_2 > M_2$.

Proof: See [8], in which a proof valid for both the single-parameter and the multiple-parameter cases is outlined.

The Kawaguchi extensor of range α : 0 to M derived from the tensor rank of the one-parameter extensor $F_{;\alpha a}$ is defined as $\binom{M}{\alpha} T_a^{\cdot(M-\alpha)}$, where T_a is a nonvanishing tensor rank of $F_{;\alpha a}$. Unless $F_{;Ma} \equiv 0$, the extensor is $\binom{M}{\alpha} F_{;Ma}^{\cdot(M-\alpha)}$. If $F_{;Ma} \equiv 0$ and $F_{;(M-1)a} \not\equiv 0$, the Kawaguchi extensor is $\binom{M-1}{\alpha} F_{;(M-1-\alpha)a}^{(M-1-\alpha)}$. For the two-parameter case, if $F_{;(M_1,M_2)a} \not\equiv 0$ for each (M_1, M_2) such that $M_1 + M_2 = M$, then there are M+1 Kawaguchi extensors, namely $\binom{Q}{\alpha_1}\binom{M-Q}{\alpha_2} F_{;(Q,M-Q)a}^{(Q-\alpha_1, M-Q-\alpha_2)}$, $Q = 0, 1, \ldots, M$, one for each tensor rank $F_{;(M_1,M_2)a}$, $M_1 + M_2 = M$, of $F_{;(\alpha_1,\alpha_2)a}$. If $F_{;(Q,M-Q)a} \equiv 0$ for each Q such that $0 \leq Q \leq M$, and if $F_{;(Q,M-1-Q)a} \not\equiv 0$ for each Q such that $0 \leq Q \leq M - 1$, there are

M Kawaguchi extensors, namely $\binom{Q}{\alpha_1}\binom{M-1-Q}{\alpha_2}F^{(Q-\alpha_1, M-1-Q-\alpha_2)}_{;(Q,M-1-Q)a}$

$Q = 0, 1, \ldots, M-1.$

7. Extensor quotient law. The quotient law for contraction over a reduced range is illustrated by the following special case, which applies to the case of two parameters as well:

Theorem (7.1). If the quantities $E_{\alpha a}$ are independent of the choice of $V^{\alpha a}$, an arbitrary excontravariant extensor, and if the quantities

$$\sum_{\alpha=B}^{M} \binom{\alpha}{B} E^{\gamma c}_{\alpha a} V^{\alpha-\beta a}$$

are the components of an extensor of excontravariant order unity, with $V^{\alpha-\beta a}$ sufficiently arbitrary that in any one given coordinate system we may assign it the values $\delta^a_d\, \delta^\alpha_\lambda$ (d and λ fixed and arbitrary except that $\lambda \geq B$), then $E^{\gamma c}_{\alpha a}$ is an extensor of excontravariant order unity and excovariant order unity of the reduced range $B \leq \alpha \leq M$.

Proof: We have given that

$$\sum_{\alpha=B}^{M} \binom{\alpha}{B} E^{\gamma c}_{\alpha a} V^{\alpha-\beta a} = \left[\sum_{\rho=B}^{M} \binom{\rho}{B} E^{\tau t}_{\rho r} V^{\rho-Br}\right] X^{\gamma c}_{\tau t} .$$

Hence upon replacing $V^{\rho-Br}$ with $\sum_{\alpha=B}^{M} V^{\alpha-Ba} X^{(\rho-B)r}_{(\alpha-B)a}$ we may write

$$\sum_{\alpha=B}^{M} V^{\alpha-Ba}\left[\binom{\alpha}{B} E^{\gamma c}_{\alpha a} - \sum_{\rho=B}^{M}\binom{\rho}{B} E^{\tau t}_{\rho r} X^{\gamma c}_{\tau t} X^{(\rho-B)r}_{(\alpha-B)a}\right] = 0.$$

If we take $V^{\alpha-\beta a}$ to be $\delta^{a}_{d}\,\delta^{\alpha}_{\lambda}$ in the x-system for a particular d and λ, $\lambda \geq B$, we have that the bracket vanishes for $a = d$ and $\alpha = \lambda$. But d and λ were arbitrary except that $\lambda \geq B$, so the bracket must vanish for all values of a and α, $\alpha \geq B$. Now by using the formula $\binom{A}{B}^{-1}\binom{\rho}{B} X^{(\rho-B)r}_{(A-B)a} = \binom{\rho}{A} X^{r(\rho-A)}_{a} = X^{\rho r}_{Aa}$, $A = \alpha$, we find that the vanishing of the bracket is equivalent to the transformation equation for an extensor of the reduced range $B \leq \alpha \leq M$:

$$E^{\gamma c}_{\alpha a} = \sum_{\rho=B}^{M} E^{\tau t}_{\rho r} X^{\gamma c}_{\tau t} X^{\rho r}_{\alpha a} \quad .$$

8. <u>Algebraic linear Vector spaces</u>. The following definitions and theorems describe algebraic linear Vector spaces:

Definition (8.1). A set S of elements is called an <u>algebraic linear Vector space</u> and its elements are called <u>Vectors</u> if S admits the two operations of addition and scalar multiplication subject to the following conditions:

Addition satisfies the five postulates:

(1) To every ordered pair (x,y) of Vectors there corresponds a uniquely defined element $x+y$ of S, known as the sum of x and y,

(2) $x + y = y + x$,

(3) $(x + y) + z = x + (y + z)$,

(4) $x + z = y + z$ implies $x = y$,

(5) For every pair (x,y) of elements in S there is an element z in S such that $x + z = y$.

Scalar multiplication satisfies:

(6) For every scalar a and every element $x \in S$, there is a uniquely defined scalar product $ax = xa$ in S.

(7) $(a + b)x = ax + bx$,

(8) $a(x + y) = ax + ay$,

(9) $a(bx) = (ab)x$,

(10) $1 \cdot x = x$.

Definition (8.2). A set $\{x_1, x_2, \ . \ . \ . \ , x_n\}$ of n Vectors is said to be a <u>linearly dependent set</u> if and only if there exists a set

$\{a_1, a_2, \ldots, a_n\}$ of n scalars containing a member different from zero such that $\sum_{i=1}^{M} a_i x_i = 0$; otherwise it is said to be a <u>linearly independent set</u>.

Definition (8.3). If a collection M of Vectors of a linear Vector space S is such that if $x \in M$ and $y \in M$, then $ax + by \in M$ for all scalars a and b, then M is said to be a <u>linear manifold</u> of S.

Definition (8.4). A set $\{e_1, e_2, \ldots, e_K\}$ of Vectors is said to <u>span</u> a linear manifold M if every Vector x in M can be represented as a linear combination of $e_1, e_2, \ldots, e_K$.

Definition (8.5). A set $\{e_1, e_2, \ldots, e_K\}$ of Vectors is said to be a <u>basis</u> for a linear manifold M if the set $\{e_1, e_2, \ldots, e_K\}$ spans M and, furthermore, is a linearly independent set.

Definition (8.6). A linear manifold M is said to be an <u>N-dimensional linear manifold</u> if a basis for M consists of N Vectors.

Remark: It may be shown that a basis for an N-dimensional linear manifold M contains exactly N Vectors. Further, any linearly independent set of N Vectors of M constitutes a basis for M. If

$B_N = \{e_1, e_2, \ldots, e_N\}$ is such a basis, and $x \in M$, we may write uniquely that

$$x = \sum_{i=1}^{N} x_i e_i ,$$

and state that the N-tuple $(x_1, x_2, \ldots, x_N)$ is the set of scalar components of x with respect to the basis B_N.

Theorems (8.1) - (8.3) are well-known and are stated without proof:

Theorem (8.1). Any subset of a linearly independent set of Vectors is a linearly independent set of Vectors.

Theorem (8.2). Every set of Vectors contains a linearly independent set.

Theorem (8.3). If A is a linearly dependent set of N distinct nonzero Vectors, then there exists a unique $K < N$ such that every subset B of A consisting of K elements (1) is linearly independent, and (2) has the property that if $a \in A$, $a \notin B$, then $B \cup \{a\}$ is linearly dependent.

Remark: Accordingly, we have the following definition.

Definition (8.7). If A is a linearly dependent set of N nonzero Vectors, and if B is a linearly independent subset of A such that $B \cup \{a\}$ is a linearly dependent set for every $a \in A$, $a \notin B$, then B is called a <u>maximal linearly independent subset of A</u>.

Remark: We introduce one further definition to augment our vocabulary concerning linear arrays.

Definition (8.8). If there are defined operations of addition (+) and scalar multiplication (·) on a subset $\mathcal{M}$ of elements of a set $\mathcal{S}$ (not necessarily an algebraic linear Vector space) such that $a \cdot x + b \cdot y \in \mathcal{M}$ for all scalars a and b whenever $x \in \mathcal{M}$ and $y \in \mathcal{M}$, then $\mathcal{M}$ is said to be a <u>subrealm</u> of the set $\mathcal{S}$.

Comment: We will thus be able to use the term 'subrealm' for a subset $\mathcal{M}$ of a set $\mathcal{S}$ -- whenever $\mathcal{M}$ is an algebraic linear Vector space under the operations in $\mathcal{M}$ of addition and scalar multiplication-- without implying that $\mathcal{S}$ is an algebraic linear Vector space under these operations. The term 'linear manifold' is reserved to carry this latter implication.

9. Parameterized arcs and hypersurfaces. Definition (9.1). If $x^a(t)$, $a = 1, 2, \ldots, N$, are functions of class c^M in the interval $I = \{t \mid t_1 \leq t \leq t_2\}$, if $\sum_{a=1}^{N} (x'^a)^2 > 0$ for every $t \in I$, and if the set $\mathcal{P} = \{P(t) = (x^1(t), x^2(t), \ldots, x^N(t)) \mid t \in I\}$ has the property that if $t_1 \in I$ and $t_2 \in I$ but $t_1 \neq t_2$, then $P(t_1) \neq P(t_2)$, then the set $\mathcal{P}$ together with the correspondences $x^a = x^a(t)$, $a = 1, 2, \ldots, N$, is said to constitute a class c^M parameterized arc in $(x^1, x^2, \ldots, x^N)$-space, with curve parameter t.

Definition (9.2). Let E_N be an N-dimensional Euclidean space having a set of orthonormal base vectors I^a and the associated rectangular Cartesian coordinates x^a, a : 1 to N; and let $\rho = x^a I_a$, $I_a = I^a$. Also, let $x^a(u^1, \ldots, u^{N-1})$ be a set of functions defined and of class c^M over a connected region R of an auxiliary u-space. Let the set $\mathcal{P}$ of vertices of $\rho(u)$, $u \in R$, have the properties: (1) If $u_1 \in R$ and $u_2 \in R$, but $u_1 \neq u_2$, then $\rho(u_1) \neq \rho(u_2)$; (2) the vector

$$e_{i_1 i_2 \cdots i_N} \frac{\partial x^{i_1}}{\partial u^1} \frac{\partial x^{i_2}}{\partial u^2} \cdots \frac{\partial x^{i_{N-1}}}{\partial u^{N-1}} I^{i_N}$$

is nonzero for every $u \in R$. Then the set $\mathcal{P}$ together with the correspondences $\rho = \rho(u^1, u^2, \ldots, u^{N-1})$ is said to constitute a class c^M parameterized hypersurface in $(x^1, x^2, \ldots, x^N)$-space with hypersurface parameters $u^1, u^2, \ldots, u^{N-1}$.

Remark: Here the symbol $e_{i_1 i_2 \ldots i_N}$ is defined for all cases such that, for each $r = 1, 2, \ldots, N$, i_r is an integer such that $1 \leq i \leq N$. The value assigned $e_{i_1 i_2 \ldots i_N}$ is 0, +1, or -1 according as some integer appears more than once in $i_1, i_2, \ldots, i_N$; as the sequence $i_1, i_2, \ldots, i_N$ is an even permutation of the sequence 1, 2, . . . , N; as the sequence $i_1, i_2, \ldots, i_N$ is an odd permutation of the sequence 1, 2, . . . , N. Here, we note that $\frac{\partial x^j}{\partial u^k}$ is the jth component of the Gibbs vector $\frac{\partial \rho}{\partial u^k}$, and summation from 1 to N is intended on repeated indices $i_1, i_2, \ldots, i_N$.

Henceforth, the parameterized arc (or hypersurface) consisting of the set $\mathcal{P}$ and the associated correspondences between points and parameter values will be referred to simply as 'the parameterized arc (or hypersurface) $\mathcal{P}$.' Henceforth, also, when N = 3, a parameterized hypersurface will be called a 'parameterized surface.'

10. Certain imbedding theorems and the alternate two-parameter methods. Theorem (10.1). Given the excovariant extensor $E_{(\alpha_1,\alpha_2)a}$ with coupled-limit range $\alpha_1 \geq 0$, $\alpha_2 \geq 0$, $\alpha_1 + \alpha_2 \leq L$. The quantities

$$T_{(\alpha_1,\alpha_2)a} = \begin{cases} E_{(\alpha_1,\alpha_2)a} & \text{when } 0 \leq \alpha_1 \leq L,\ 0 \leq \alpha_2 \leq L,\ 0 \leq \alpha_1 + \alpha_2 \leq L \\ 0 & \text{when } 0 \leq \alpha_1 \leq L,\ 0 \leq \alpha_2 \leq L,\ L+1 \leq \alpha_1 + \alpha_2 \leq 2L \end{cases}$$

are the components of an extensor of uncoupled-limit range $0 \leq \alpha_1 \leq L$, $0 \leq \alpha_2 \leq L$ (into which $E_{(\alpha_1,\alpha_2)a}$ is imbedded).

Proof:

$$\sum_{\alpha_1=0}^{L} \sum_{\alpha_2=0}^{L} T_{\alpha a} X^{\alpha a}_{\rho r} = \sum_{\substack{\alpha_1=0 \\ 0\leq\alpha_1+\alpha_2\leq L}}^{L} \sum_{\alpha_2=0}^{L} T_{\alpha a} X^{\alpha a}_{\rho r}$$
$$0\leq\alpha_1+\alpha_2\leq M$$

$$+ \sum_{\substack{\alpha_1=0 \\ L+1\leq\alpha_1+\alpha_2\leq 2L}}^{L} \sum_{\alpha_2=0}^{L} T_{\alpha a} X^{\alpha a}_{\rho r} = \sum_{\alpha_1=0}^{L} \sum_{\alpha_2=0}^{L} E_{\alpha a} X^{\alpha a}_{\rho r}$$

$$= \begin{cases} 0 & \text{if } 0 \leq \rho_1 \leq L,\ 0 \leq \rho_2 \leq L,\ L+1 \leq \rho_1 + \rho_2 \leq 2L \\ E_{\rho r} & \text{if } 0 \leq \rho_1 \leq L,\ 0 \leq \rho_2 \leq L,\ 0 \leq \rho_1 + \rho_2 \leq L \end{cases}$$

$$= T_{(\rho_1,\rho_2)r}$$

Comment: The nonvanishing tensor ranks of $T_{(\alpha_1,\alpha_2)a}$ are the ranks for which $\alpha_1 + \alpha_2 = L$, for then $T_{(\alpha_1,\alpha_2)a} = E_{(\alpha_1,\alpha_2)a}$ $= E_{(\rho_1,\rho_2)r}X^r_a = T_{(\rho_1,\rho_2)r}X^r_a$.

Theorem (10.2). Given the excovariant extensor $E_{(\alpha_1,\alpha_2)a}$ with uncoupled-limit range $0 \leq \alpha_1 \leq M_1$, $0 \leq \alpha_2 \leq M_2$. The quantities

$$T_{(\alpha_1,\alpha_2)a} = \begin{cases} E_{(\alpha_1,\alpha_2)a} & \text{when } 0 \leq \alpha_1 \leq M_1, 0 \leq \alpha_2 \leq M_2, 0 \leq \alpha_1 + \alpha_2 \leq M_1 + M_2 \\ 0 & \text{when } \alpha_1 > M_1 \text{ or } \alpha_2 > M_2;\ 0 \leq \alpha_1 + \alpha_2 \leq M_1 + M_2 \end{cases}$$

are the components of an extensor of coupled-limit range $\alpha_1 \geq 0$, $\alpha_2 \geq 0$, $\alpha_1 + \alpha_2 \leq M_1 + M_2$ (into which $E_{(\alpha_1,\alpha_2)a}$ is imbedded).

Proof:
$$\sum_{\alpha_1+\alpha_2\geq 0}^{\alpha_1+\alpha_2\leq M_1+M_2} T_{\alpha a} X^{\alpha a}_{\rho r} = \sum_{\substack{\alpha_1+\alpha_2\geq 0 \\ 0\leq\alpha_1\leq M_1,\ 0\leq\alpha_2\leq M_2}}^{\alpha_1+\alpha_2\leq M_1+M_2} T_{\alpha a} X^{\alpha a}_{\rho r}$$

$$+ \sum_{\substack{\alpha_1+\alpha_2\geq 0 \\ \alpha_1>M_1 \text{ or } \alpha_2>M_2}}^{\alpha_1+\alpha_2\leq M_1+M_2} T_{\alpha a} X^{\alpha a}_{\rho r} = \sum_{\alpha_1=0}^{M_1} \sum_{\alpha_2=0}^{M_2} E_{\alpha a} X^{\alpha a}_{\rho r}$$

$$= \begin{cases} 0 \text{ if } \rho_1 > M_1 \text{ or } \rho_2 > M_2; 0 \leq \rho_1 + \rho_2 \leq M_1 + M_2 \\ E_{\rho r} \text{ if } 0 \leq \rho_1 \leq M_1, 0 \leq \rho_2 \leq M_2; 0 \leq \rho_1 + \rho_2 \leq M_1 + M_2 \end{cases}$$

$$= T_{(\rho_1,\rho_2)r}.$$

Comment: If $E_{(M_1,M_2)a} \not\equiv 0$, the nonvanishing tensor rank of $T_{(\alpha_1,\alpha_2)a}$ is rank $\alpha_1 = M_1$, $\alpha_2 = M_2$, for $T_{(M_1,M_2)a} = E_{(M_1,M_2)a}$ $= E_{(M_1,M_2)r} X^r_a = T_{(M_1,M_2)r} X^r_a$.

11. <u>A certain invariant as a reduced-range extensor contraction</u>. Theorem (11.1). Given an absolute excovariant one-parameter extensor $E_{\alpha a}$ of range $0 \leq \alpha \leq M$, and given an arbitrary class C^M contravariant tensor V^a. The reduced range extensor contraction

$$\sum_{\alpha=L}^{M} \binom{\alpha}{L} E_{\alpha a} V^{a(\alpha-L)} \quad ,$$

where $0 \leq L \leq M$, is an invariant.

Proof:

$$\sum_{\alpha=L}^{M} \binom{\alpha}{L} E_{\alpha a} V^{a(\alpha-L)} = \sum_{\alpha=L}^{M} \binom{\alpha}{L} \sum_{\rho=0}^{M} \overline{E}_{\rho r} X^{\rho r}_{\alpha a} \cdot \sum_{\mu=0}^{\alpha-L} \binom{\alpha-L}{\mu} \overline{V}^{s(\mu)} X^{a(\alpha-L-\mu)}_{s}$$

by the Leibnitz formula. The right member in this last expression may be written

$$\sum_{\alpha=L}^{M} \sum_{\rho=0}^{M} \sum_{\mu=0}^{\alpha-L} \binom{\alpha}{L} \overline{E}_{\rho r} \overline{V}^{s(\mu)} \binom{\alpha-L}{\mu} X^{\rho r}_{\alpha a} \binom{\alpha}{L+\mu}^{-1} X^{\alpha a}_{(L+\mu)s} \quad .$$

Because $\binom{\alpha}{L}\binom{\alpha-L}{\mu}\binom{\alpha}{L+\mu}^{-1} = \binom{L+\mu}{L}$ and because $\binom{\alpha-L}{\mu} = 0$ for $\mu > \alpha - L$ (so that the upper limit on the μ-summation may be raised to $M - L$), we may replace this last expression by

$$\sum_{\alpha=L}^{M} \sum_{\rho=0}^{M} \sum_{\mu=0}^{M-L} \binom{L+\mu}{L} \overline{E}_{\rho r} \overline{V}^{s(\mu)} X^{\rho r}_{\alpha a} X^{\alpha a}_{(L+\mu)s} .$$

By an interchange of the order of summation this last expression becomes

$$\sum_{\rho=0}^{M} \sum_{\mu=0}^{M-L} \binom{L+\mu}{L} \overline{E}_{\rho r} \overline{V}^{s(\mu)} \left[\sum_{\alpha=L}^{M} X^{\rho r}_{\alpha a} X^{\alpha a}_{(L+\mu)s} \right] .$$

The bracketed portion of this expression becomes $\delta^{\rho}_{L+\mu} \delta^{r}_{s}$, by a property of the symbols $X^{\rho r}_{\alpha a}$ and $X^{\alpha a}_{(L+\mu)s}$. The preceding expression becomes

$$\sum_{\rho=0}^{M} \sum_{\mu=0}^{M-L} \binom{L+\mu}{L} \overline{E}_{\rho r} \overline{V}^{s(\mu)} \delta^{\rho}_{L+\mu} \delta^{r}_{s} ,$$

or

$$\sum_{\rho=L}^{M} \binom{\rho}{L} \overline{E}_{\rho r} \overline{V}^{r(\rho-L)} ,$$

and the proof is completed.

Theorem (11.2). (a two-parameter version of Theorem (11.1)). Given a two-parameter excovariant extensor $E_{(\alpha_1,\alpha_2)a}$ of coupled-limit range $\alpha_1 \geq 0$, $\alpha_2 \geq 0$, $\alpha_1 + \alpha_2 \leq M$, and given an arbitrary class c^M

contravariant tensor V^a. If $L_1 \geq 0$, $L_2 \geq 0$, $L_1 + L_2 \leq M$, and if $V^{a(\beta_1,\beta_2)} \overset{def}{\equiv} 0$ when either $\beta_1 < 0$ or $\beta_2 < 0$, then the reduced range contraction

$$\sum_{\alpha_1+\alpha_2 \geq L_1+L_2}^{\alpha_1+\alpha_2 \leq M} \binom{\alpha_1}{L_1}\binom{\alpha_2}{L_2} E_{(\alpha_1,\alpha_2)a} V^{a(\alpha_1-L_1,\alpha_2-L_2)}$$

is an invariant.

Proof (Outline): First, prove a lemma regarding the invariance of the combination of a two-parameter excovariant extensor $T_{(\alpha_1,\alpha_2)a}$ of uncoupled-limit range $0 \leq \alpha_1 \leq M$, $0 \leq \alpha_2 \leq M$, with an arbitrary class c^{2M} contravariant tensor V^a according to the manner

$$\sum_{\alpha_1=0}^{M} \sum_{\alpha_2=0}^{M} \binom{\alpha_1}{L_1}\binom{\alpha_2}{L_2} T_{(\alpha_1,\alpha_2)a} V^{a(\alpha_1-L_1,\alpha_2-L_2)}$$

with $0 \leq L_1 \leq M$, $0 \leq L_2 \leq M$. This proof is formally like the proof for the one-parameter Theorem (11.1), with M replaced by (M,M). In particular we have the invariance of this last expression whenever $T_{(\alpha_1,\alpha_2)a}$ has a coupled-limit extensor $E_{(\alpha_1,\alpha_2)a}$ imbedded in its lower ranks, with the higher ranks of $T_{(\alpha_1,\alpha_2)a}$ vanishing, i.e., whenever the quantities

$$T_{(\alpha_1,\alpha_2)a} = \begin{cases} E_{(\alpha_1,\alpha_2)a} & \text{when } 0 \le \alpha_1 \le M,\ 0 \le \alpha_2 \le M,\ 0 \le \alpha_1 + \alpha_2 \le M \\ 0 & \text{when } 0 \le \alpha_1 \le M,\ 0 \le \alpha_2 \le M,\ M+1 \le \alpha_1 + \alpha_2 \le 2M \end{cases}$$

are the components of an extensor of range $0 \le \alpha_1 \le M$, $0 \le \alpha_2 \le M$. But then the invariant with iterated summation on α_1 and α_2 can be written as the coupled-limit summation

$$\sum_{\alpha_1+\alpha_2\ge 0}^{\alpha_1+\alpha_2\le M} \binom{\alpha_1}{L_1}\binom{\alpha_2}{L_2} E_{(\alpha_1,\alpha_2)a} V^{a(\alpha_1-L_1,\alpha_2-L_2)}$$

and the theorem is proved.

12. Summary of certain elementary concepts of the calculus of variations. From Section 9 we note that a parameterized arc is a different entity from the collection of points which forms the associated curve. One point of distinction is that the relations $x^a = x^a(t)$ not only determine a curve, but also order the points of the arc. Thus, although a parameter transformation may not change the curve, it does, excluding the identity transformation, change the parameterized arc.

Definition (12.1). A parameter transformation $t = t(T)$, $T_1 \le T \le T_2$, will be said to be admissible of class C^M if the function

$t(T)$ is single-valued and of class c^M and such that $\frac{dt}{dT}$ is positive for each value of T such that $T_1 \leq T \leq T_2$.

Definition (12.2). A parameterized arc $x^a = x^a(t)$, $t_1 \leq t \leq t_2$, is said to be of class c^M, $M > 0$, if $x^a(t)$ are class c^M functions and if $\sum_{a=1}^{N} (x'^a)^2$ is positive for every t such that $t_1 \leq t \leq t_2$.

Comment: We may now proceed to approach the subject of the calculus of variations by introducing certain functions from a parameterized arc to a real number. For purposes of the calculus of variations we will wish to treat the requirement that such functions be independent of the choice of the parameter.

For example, we may consider a function $F(x, x')$ of the coordinate variables and their derivatives with respect to a curve parameter t. If C is a parameterized arc, given by $x^a = x^a(t)$, of class c' for t in the interval $t_1 \leq t \leq t_2$, then substitution of $x^a(t)$ and $x'^a(t)$ for each x and x' in F will yield a function of t, thereby determining a value of F at each point of C. If this is an integrable function, then there is determined a value for

$I = \int_{t_1}^{t_2} F(x, x')dt$, and we may say that this integral is a function from a class c' parameterized arc to a real number. To examine the requirement that I be independent of the choice of parameter t, we presume the following theorem:

Theorem (12.1). A necessary and sufficient condition that the integral from the fixed point P to the fixed point Q of the continuous function $F(x, x')$ along any class c' parameterized arc (such that the quantities $x^a, x^{a'}$ associated with C are in the region of continuity of F) transform according to the equation

$$\int_{t(P)}^{t(Q)} F(x,x')dt = \int_{T(P)}^{T(Q)} F\left(X,\frac{dX}{dT}\right) dT$$

whenever t undergoes an admissible class c' parameter transformation $t = t(T)$ is that

$$F(x,kx') = kF(x,x')$$

hold identically in x,x' and all positive values of k.

Comment: By differentiation of $F(x,kx') = kF(x,x')$ with respect to k and evaluation at $k = 1$, we obtain an alternate necessary and sufficient condition for the invariance of $I = \int_{t_1}^{t_2} F(x,x')dt$ with respect to admissible parameter transformations, namely

$$(12.1) \qquad F \equiv x'^a F_{;1a} \quad .$$

By differentiation of (12.1) with respect to t, we obtain

$$(12.2) \qquad \sum_{\alpha=0}^{1} x'^{a(\alpha)} F_{\alpha a} = \sum_{\alpha=0}^{1} x'^{a(\alpha)} F_{;\alpha a}$$

where $F_{;\alpha a}$ is the gradient extensor of Section 6 associated with the class c' invariant function $F(x,x')$, and where $F_{\alpha a} \left(\equiv \binom{1}{A} F_{;1a}{}^{\cdot(1-A)}, A=\alpha\right)$ is the associated Kawaguchi extensor. Here we assume $F_{;1a} \neq 0$. These two extensors are called the primary extensors associated with $F(x,x')$. Now because x'^a is a contravariant tensor, $x'^{a(\alpha)}$ is an excontravariant extensor. Accordingly, equation (12.2) asserts the equality of two quantities which are extensor contractions and therefore invariants. It says that the primary extensors associated with $F(x,x')$ must have equal contractions with the extensor $x'^{(\alpha)}$.

With regard to the simple calculus of variations problem based upon the function $I(C) = \int_{t_1}^{t_2} F(x,x')dt$, we first introduce a family of comparison curves, defined by a base arc C_0, a set of variation functions $V^a(t)$, and a variation parameter v.

Specifically, we require that the function $F(x,x')$ be of class c'' for the x's in a connected region R and for all values of the x''s excepting for the excluded set $x'^1 = x'^2 = \ldots = x'^N = 0$. We further require that $F(x,x')$ satisfy (12.2). Next, we assume that we are given a pair of fixed points P and Q in R and a class c'' parameterized arc $C_0 : x^a = x_0^a(t)$ which passes through P and Q for $t = t_1$ and $t = t_2$, respectively, with $t_1 < t_2$. Finally, the variation functions $V^a(t)$ are assumed to be of class c'' and to satisfy $V^a(t_1) = V^a(t_2) = 0$.

It follows that there is a number interval i(v) containing zero such that, for each fixed value of v in i(v), the equations

$$(12.3) \qquad x^a = x^a(t,v) = x_0^a(t) + vV^a(t)$$

define a parameterized arc $C(v)$ which is an admissible comparison curve. In particular $C(0)$ is the base arc C_0.

If we now replace each x in $F(x,x')$ by $x^a(t,v)$, and each x' in $F(x,x')$ by $\frac{\partial x^a(t,v)}{\partial t} = x'^a(t) + vV'^a(t)$, then $F(x,x')$ by virtue of this substitution becomes a function of t and v, of class c'' in v. The integral $I = \int_{t_1}^{t_2} F(x,x')dt$ is now a class c'' function of v alone, which we denote by $I(v)$ when taken along $C(v)$.

To compute $\left.\frac{dI}{dv}\right|_{v=0}$ we note (1) because the limits t_1 and t_2 on the integral are independent of v and because F is a class c'' function of v, we may pass over the integral sign and differentiate F; (2) if we replace the x's in $F(x,x')$ by the functions $x^a(t,v)$, select a value t_0 in the interval $t_1 \leq t \leq t_2$, and assign v the value zero, then the partial derivative of F with respect to the symbol $x^a(t,v)$ evaluated at $x^a(t_0,0)$ is the same as the partial derivative of F in its original form with respect to x at $x^a = x_0^a(t_0)$ since $x^a(t_0,0) = x_0^a(t_0)$. Hence to evaluate $\left.\frac{dI}{dv}\right|_{v=0}$ we pass over the integral sign and (1) compute $\frac{\partial F(x,x')}{\partial x^a}$, (2) multiply the result by $\frac{\partial x^a}{\partial v}\left(= V^a(t)\right)$, (3) compute

$\frac{\partial F(x,x')}{\partial x'^a}$ and, (4) multiply by $\frac{\partial x'^a}{\partial v}\left(= V^{a\prime}(t)\right)$, (5) add the products of (2) and (4). In this way we obtain

$$(12.4) \qquad \left.\frac{dI}{dv}\right|_{v=0} = \int_{t_1}^{t_2} \sum_{\alpha=0}^{1} F_{;\alpha a} V^{a(\alpha)} dt = \int_{t_1}^{t_2} \left(F_{;0a} V^a + F_{;1a} V^{a\prime} \right) dt \ .$$

Now a necessary condition that the arc C_0 must satisfy in order to furnish an extreme value to $I = \int_{t_1}^{t_2} F(x,x')dt$ relative to the set of arcs which are admissible comparison curves is that $\left.\frac{dI}{dv}\right|_{v=0} = 0$. Thus the first variation of the integral I must vanish when C_0 is the base arc.

The classical procedure for determining the Euler differential equations which describe such extremizing arcs is to integrate one of the terms in equation (12.4) by parts, then prove a theorem called the fundamental lemma which enables one to discard the variation functions V^a and the integral sign and thereby obtain the Euler differential equations. The following theorem constitutes the fundamental lemma of the calculus of variations and is stated without proof:

Theorem (12.2). If F(t) is continuous at each point of the closed interval $t_1 \leq t \leq t_2$ and if for each function $\xi(t)$ of class c''

such that $\xi(t_1) = \xi(t_2) = 0$ we have $\int_{t_1}^{t_2} F(t)\ \xi(t)\ dt$ vanishes, then $F(t) \equiv 0$ on the interval $t_1 \leq t \leq t_2$.

Comment: A number of facts involving the primary extensors $F_{;\alpha a}$ and $F_{\alpha a}$ suggest the value of a more heuristic approach involving extensors, rather than integration by parts and the fundamental lemma. These facts are as follows:

(1) The differential equations sought should be of order two or more, because they must have a solution passing through two fixed points which might be selected in infinitely many ways. The equation $F_{;\alpha a} - F_{\alpha a} = 0$ is of second order except when $F_{;1a;1b} = 0$.

(2) The extensors $F_{;\alpha a}$ and $F_{\alpha a}$ already agree in their tensor rank and, because of the imposed invariance of $\int_{t_1}^{t_2} F(x,x')dt$ under admissible parameter transformation, were seen in (12.2) to have equal contractions with the extensor $x'^{a(\alpha)}$.

(3) The value of $\left.\frac{dI}{dv}\right|_{v=0}$ for a given base arc C_0 is equal to the integral of the extensor contraction $\sum_{\alpha=0}^{1} F_{;\alpha a}V^{a(\alpha)}$ and is therefore invariant under coordinate transformations. Consequently, the curves

C_0 determined by $\left.\frac{dI}{dv}\right|_{v=0} = 0$ are independent of the coordinate system. But if $F_{;\alpha a} = F_{\alpha a}$ is satisfied in coordinate system (x) by the curve C_0, then the mate equation $\overline{F}_{;\rho r} = \overline{F}_{\rho r}$ is satisfied in coordinate system $(\overline{x})$ by the same curve C_0.

(4) The equality of $F_{;\alpha a}$ and $F_{\alpha a}$ converts the first variation of $I = \int_{t_1}^{t_2} F(x,x')dt$ into the integral of a derivative and, accordingly, causes it to vanish. Specifically,

$$(12.5)\quad \left.\frac{dI}{dv}\right|_{v=0} = \int_{t_1}^{t_2} \sum_{\alpha=0}^{1} F_{\alpha a}V^{a(\alpha)}dt = \int_{t_1}^{t_2} (F_{;1a}V^a)'dt = F_{;1a}V^a\Big|_{t_1}^{t_2} = 0.$$

Comment: These facts strongly suggest the extensor structure of the Euler differential equations associated with $F(x,x')$. They also suggest the equivalence of the set of Euler equations to the equality of the primary extensors associated with $F(x,x')$.

In fact we see that equality of $F_{;\alpha a}$ and $F_{\alpha a}$ along an arc is equivalent to $F_{;\alpha a} - F_{\alpha a} = 0$, whose tensor rank $\alpha = 1$ vanishes identically. The next removed rank $\alpha = 0$ becomes a tensor rank, i.e.,

$F_{;0a} - F_{0a} = 0$, or

$$(12.6)\quad F_{;0a} - F_{;1a}{}' = 0 \quad ,$$

which are the Euler equations associated with $\int_{t_1}^{t_2} F(x,x')dt$.

Remark: The related simple isoperimetric problem of the calculus of variations in which one seeks a necessary condition for an extreme value of $\int_{t_1}^{t_2} F(x,x')dt$ subject to the constraint $\int_{t_1}^{t_2} G(x,x')dt = \text{const}$ may be approached by assuming a linear relationship among the primary extensors associated with $F(x,x')$ and $G(x,x')$. The coefficients are chosen so as to guarantee the emergence of second order differential equations, and so as to avoid overdetermination. The Lagrangian multiplier rule emerges as a natural consequence in that the linear extensor equation arrived at is equivalent to the set of Euler equations associated with $F + \lambda G$, λ a Lagrangian multiplier.

For the double integral problem $\iint_{D_0} F\left(x, x^{(0,1)}, x^{(1,0)}\right) du^1 du^2$ with $N = 3$ the concept of 'comparison curve' is replaced with that of 'comparison surface.' These parameterized surfaces $x^a = x^a(u^1,u^2)$ are defined over a domain D in an auxiliary u^1,u^2-plane and consist of a closed curve $C : u^1 = u^1(t),\ u^2 = u^2(t)$ plus its interior. The surfaces are all required to pass through the same closed curve C_0 given by $x^a = x^a\left[u^1(t),\ u^2(t)\right]$.

The linear extensor equation which sets the two-parameter gradient extensor $F_{;\alpha a}$ associated with F equal to the sum of the two associated two-parameter Kawaguchi extensors is seen to make $\left.\frac{dI}{dv}\right|_{v=0} = 0$ by converting $\left.\frac{dI}{dv}\right|_{v=0}$ into the integral of a derivative. This linear extensor equation is seen to be equivalent to the appropriate Euler equations

$$(12.7) \qquad F_{;(0,0)a} - F_{;(0,1)a}{}^{(0,1)} - F_{;(1,0)a}{}^{(1,0)} = 0 \ .$$

We may now consider a simple calculus of variations problem in which the arguments of the integrand function involve higher parameter derivatives of the coordinate variables, formulated briefly as follows: Given: (1) a function $F(x, x', \ldots, x^{(M)})$ of class C^{M+1} for x in a connected region R and for all values of the x-primes excepting the excluded set $x'^1 = x'^2 = \ldots = x'^N = 0$, (2) a pair of points P, Q in R, (3) the set S of all arcs $C : x^a = x^a(t)$ of class C^{2M} which pass through P for $t = t_1$ and through Q for $t = t_2$, lie entirely in R, and satisfy the boundary conditions $x^{(\alpha)a}(t_1) = x_1^{\alpha a}$, $x^{(\alpha)a}(t_2) = x_2^{\alpha a}$ for $\alpha = 0, 1, \ldots, M-1$. Find an admissible arc $C_0 : x^a = x_0^a(t)$ in the subset S_v of S, where S_v consists of comparison curves $C_v : x^a = x^a(t,v) = x_0^a(t) + vV^a(t)$, such that C_0 causes the first variation of $\int_{t_1}^{t_2} F(x,x', \ldots, x^{(M)})dt$ to vanish relative to S_v. Here, v is a variation parameter and $V^a(t)$ are class c^M variation functions such that $V^{a(\alpha)}(t_1) = V^{a(\alpha)}(t_2) = 0$ for $\alpha = 0, 1, \ldots, M-1$.

With relation to this problem it has been noted in [19] that it is possible to generalize the concept of primary extensor so as to apply to a class c^{M+1} invariant function $F(x,x', \ldots, x^{(M)})$ evaluated over a class c^{2M} parameterized arc. This involved constructing from $F(x,x', \ldots, x^{(M)})$ not only (1) a gradient extensor $0_{\alpha a}(\equiv F_{;\alpha a})$, but also (2) a set of M additional extensors $L_{\alpha a}$ (with $L = 1, 2, \ldots, M$) whose ranks also are composed of parameter derivatives of the quantities $F_{;0a}, F_{;1a}, \ldots, F_{;Ma}$, such that $L_{0a} = F_{;La}^{(L)}$ for each L such that $0 \leq L \leq M$. The quantities $L_{\alpha a}$ for each L such that $0 \leq L \leq M$ were defined by

$$(12.8) \qquad L_{\alpha a} = \sum_{\mu=0}^{L} \binom{L}{\mu}\binom{A+\mu}{L} F_{;(A+\mu)a}^{(\mu)}, \; A = \alpha.$$

Next, it was shown that each of the M extensors $L_{\alpha a}$, $L = 1, 2, \ldots, M$, forms a perfect derivative when contracted with $V^{a(\alpha)}$, where V^a is an arbitrary class c^M tensor. Specifically, we observed that

$$(12.9)\qquad \sum_{\alpha=0}^{M} L_{\alpha a} V^{a(\alpha)} = \left[\sum_{\gamma=L}^{M} \binom{\gamma}{L} F_{;\gamma a} V^{a(\gamma-L)}\right]^{(L)}$$

for L = 1, 2, . . . , M.

Accordingly, it was noted that the linear extensor equation

$$(12.10)\qquad \sum_{L=0}^{M} (-1)^L \; L_{\alpha a} = 0$$

converts the first variation to the integral of a derivative, so that the first variation $\left.\frac{dI}{dv}\right|_{v=0}$ vanishes. In particular there was obtained

$$(12.11)\qquad \left.\frac{dI(v)}{dv}\right|_{v=0} = \int_{t_1}^{t_2} \sum_{\alpha=0}^{M} F_{;\alpha a} V^{a(\alpha)} dt = - \int_{t_1}^{t_2} \sum_{\alpha=0}^{M} \sum_{L=1}^{M} (-1)^L L_{\alpha a} V^{a(\alpha)} dt$$

$$= \left.\left[\sum_{L=1}^{M} (-1)^{L+1} \sum_{\gamma=L}^{M} \binom{\gamma}{L} F_{;\gamma a} V^{a(\gamma-L)}\right]^{(L-1)}\right|_{t_1}^{t_2} = 0 .$$

Furthermore, each nonzero rank of (12.10) vanished identically. Therefore, rank 0 was seen to be a tensor rank, and in particular to express the Euler equations associated with $\int_{t_1}^{t_2} F(x,x', . . . , x^{(M)})dt$, namely

$$(12.12)\qquad \sum_{\alpha=0}^{M} (-1)^{\alpha} F_{;\alpha a}{}^{(\alpha)} = 0 \ .$$

A similar treatment of the isoperimetric problem of the type $\int_{t_1}^{t_2} F(x,x', \ldots, x^{(M)})dt$, $\int_{t_1}^{t_2} G(x,x', \ldots, x^{(M)})dt = \text{const}$, reveals that the appropriate Euler equations may be expressed as a linear equation relating $F_{;\alpha a}$, $G_{;\alpha a}$, and the 2M additional extensors whose zero ranks, respectively, are $F_{;0a}$, $F_{;1a}{}'$, . . . , $F_{;Ma}{}^{(M)}$; $G_{;0a}$, $G_{;1a}{}'$, . . . , $G_{;Ma}{}^{(M)}$. There is also an extensor solution to the surface integral problem of the type $\iint_{D_0} F(x,x^{(0,1)}, x^{(1,0)}, \ldots, x^{(M_1,M_2)})du^1du^2$.

13. <u>The set of Synge tensors associated with $E_{\alpha a}$ and the variational integrals $\int_{t_1}^{t_2} \sum_{\alpha=0}^{M} E_{\alpha a}V^{a(\alpha)}dt$, $\iint_{D_0} \sum_{\alpha_1+\alpha_2\geq 0}^{\alpha_1+\alpha_2\leq M} E_{\alpha a} V^{a(\alpha)}du^1du^2$.</u>

A certain set of tensors associated with the class c^{M+1} invariant function $F(x,x', \ldots, x^{(M)})$ was given by Synge in [17]. A simple generalization consists of obtaining a related set S_s of tensors $S_a^{[0]}(E_{\alpha a})$, $S_a^{[1]}(E_{\alpha a})$, . . . , $S_a^{[M]}(E_{\alpha a})$, where

$$(13.1)\qquad S_a^{[\gamma]}(E_{\alpha a}) = \sum_{\alpha=0}^{M-\gamma} (-1)^{\alpha} \binom{\alpha+\gamma}{\gamma} E_{(\alpha+\gamma)a}{}^{(\alpha)}$$

for $\gamma = 0, 1, \ldots, M$. We will refer to $S_s = \{S_a^{[\gamma]}(E_{\alpha a}) \mid \gamma = 0, 1, \ldots, M\}$ as the set of Synge tensors associated with $E_{\alpha a}$. Our purpose in this section is to give a simple interpretation with respect to certain integrals of the calculus of variations for each Synge tensor in S_s. Consideration of a further variational role for the tensors in S_s is postponed until Section 31.

Accordingly, we will recall that the vanishing of the Synge tensor $S_a^{[0]}(E_{\alpha a})$ was seen in [15] to describe a normalizing arc relative to an extensor-generalized calculus of variations problem involving the integral $\int_{t_1}^{t_2} \sum_{\alpha=0}^{M} E_{\alpha a} V^{a(\alpha)} dt$. We will further note from [15] that, in case $E_{\alpha a} = F_{;\alpha a}$, the vanishing of $S_a^{[0]}(E_{\alpha a})$ expresses the Euler differential equations which describe stationary arcs in an ordinary simple calculus of variations problem involving the integral $\int_{t_1}^{t_2} F(x, x', \ldots x^{(M)}) dt$. We will then observe that a simple alteration of the equation on line 4, page 272 of [6] allows us to express a condition for the invariance with respect to admissible parameter transformations of $\int_{t_1}^{t_2} F(x, x', \ldots, x^{(M)}) dt$ as the equality of F with a certain sum of derivatives of tensor contractions involving

$S_a^{[1]}(F_{;\alpha a})$, $S_a^{[2]}(F_{;\alpha a})$, . . . , $S_a^{[M]}(F_{;\alpha a})$. We will conclude by

showing that (1) Synge tensors are involved in a certain extensor-

generalized variational problem, and (2) two-parameter Synge tensors

are involved in a certain extensor-generalized double integral varia-

tional problem.

Let us begin by reproducing in condensed form the formulation

of [15] of an extensor generalization of a simple calculus of varia-

tions problem: Suppose that there is given

(1) an extensor $E_{\alpha a}$ of range α : 0 to M whose components

are functions of $x, x', . . . , x^{(P)}$ and are of class c^M for x in a

connected region R and for all values of the x-primes which satisfy

$\Sigma(x'^a)^2 > 0$,

(2) the interval $t_1 \leq t \leq t_2$ on an auxiliary t-axis,

(3) a pair of points P_1, P_2 in R with coordinates denoted

by x_1^a, x_2^a;

(4) the set of arcs $x^a = x^a(t)$, $t_1 \leq t \leq t_2$, of class c^H

which lie entirely in R, satisfy the requirement $\Sigma(x'^a)^2 > 0$ and the boundary conditions $x^{(\alpha)a}(t_1) = x_1^{\alpha a}$, $x^{(\alpha)a}(t_2) = x_2^{\alpha a}$. Here, $0 \leq \alpha \leq M-1$, and H is the functional order of the expression

$$\sum_{\alpha=0}^{M} (-1)^{\alpha} E_{\alpha a}{}^{(\alpha)} \quad .$$

The generalized problem is then stated as follows: Given items (1) - (4) find an admissible arc $C_0 : x^a = x_0^a(t)$ such that if S_v, given by $C_v : x^a = x^a(t,v)$, is a set of admissible arcs with $x^a(t,0) = x_0^a(t)$, then for $E_{\alpha a}$ evaluated along C_0,

$$(13.2) \qquad \int_{t_1}^{t_2} \sum_{\alpha=0}^{M} E_{\alpha a} V^{a(\alpha)} dt = 0,$$

where V^a is a variation function associated with the set S_v of arcs. The vanishing of the integral in (13.2) may be interpreted geometrically to mean that $E_{\alpha a}$, defined over the α-range 0 to M is orthogonal on the average to the variation extensor $V^{(\alpha)a}$ for each admissible choice of V^a in a space of N(M+1) dimensions. The solution of (13.2) gives us a first encounter with the Synge tensor $S_a^{[0]}(E_{\alpha a})$ associated with $E_{\alpha a}$ as the Hamiltonian tensor involved in the necessary and sufficient condition

$$(13.3) \qquad S_a^{[0]}(E_{\alpha a}) = 0 \quad , \quad S_a^{[0]}(E_{\alpha a}) = \sum_{\alpha=0}^{M} (-1)^{\alpha} E_{\alpha a}{}^{(\alpha)}$$

for C_0 to be a "normalizing arc" for $E_{\alpha a}$, i.e., an arc for which (13.2) is satisfied.

Further, in case $E_{\alpha a}$ is a gradient extensor $F_{;\alpha a}$ of range 0, 1, . . . , M in (13.2), then we may regard (13.2) as the first variation of the integral

$$(13.4) \qquad I = \int_{t_1}^{t_2} F(x, x', \ldots, x^{(M)})dt \ .$$

Here, not only do we encounter the tensor $S_a^{[0]}(F_{;\alpha a})$ as the tensor involved in the condition

$$(13.5) \qquad S_a^{[0]}(F_{;\alpha a}) = 0 \ , \quad S_a^{[0]}(F_{;\alpha a}) = \sum_{\alpha=0}^{M} (-1)^{\alpha} F_{;\alpha a}{}^{(\alpha)}$$

that C_0 furnish a stationary value for the integral I of (13.4), but we shall see that the tensors $S_a^{[1]}(F_{;\alpha a})$, $S_a^{[2]}(F_{;\alpha a})$, . . . , $S_a^{[M]}(F_{;\alpha a})$ occur in the condition (13.6) that the integral I of (13.4) shall be invariant with respect to arbitrary admissible parameter transformations.

$$(13.6)\qquad F = \sum_{\gamma=1}^{M} \gamma \left[S_a^{[\gamma]}(F_{;\alpha a})x'^{a}\right]^{(\gamma-1)}.$$

Again, our point of primary interest is that the Synge tensors associated with an absolute excovariant extensor $E_{\alpha a}$ have a role with regard to a calculus of variations integral, at least when (13.4) is given, so that $E_{\alpha a} = F_{;\alpha a}$. Hence, we have noted the Eulerian structure and role of $S_a^{[0]}(E_{\alpha a})$, and we have noted the appearance of the remaining Synge tensors in a condition for the invariance of (13.4) with respect to admissible parameter transformations. We will anticipate a modified role for the set of Synge tensors in other calculus of variations problems considered in Chapter VI.

Also of interest is the result that a similar treatment of a simple generalized isoperimetric problem whose formulation we indicate briefly by

$$\int_{t_1}^{t_2} \sum_{\alpha=0}^{M} E_{\alpha a} V^{a(\alpha)} dt, \quad \int_{t_1}^{t_2} \sum_{\alpha=0}^{M} T_{\alpha a} V^{a(\alpha)} dt = 0$$

reveals that a normalizing arc should be regarded as one for which $S_a^{[0]}(P_{\alpha a})$ is the Hamiltonian tensor associated with the extensor

(13.7) $\quad P_{\alpha a} = E_{\alpha a} + \lambda T_{\alpha a}$

where λ is a Lagrangian multiplier.

Let us now consider an analogous two-parameter generalized simple variational problem, formulated as follows: Assume there is given: (1) a two-parameter extensor $E_{\alpha a}$ of range $0 \leq \alpha_1 + \alpha_2 \leq M$ whose components are functions of x, $x^{(0,1)}$, $x^{(1,0)}$, . . . , $x^{(0,P)}$, . . . , $x^{(P,0)}$ and are of class c^M for x in a connected region R and for all values of the x-primes which satisfy $\Sigma\left[x^{(0,1)a}\right]^2 > 0$ and $\Sigma\left[x^{(1,0)a}\right]^2 > 0$; (2) the domain D in an auxiliary u^1,u^2-plane, with boundary curve C; (3) a closed space curve C_0, the map of C by $x^a = x^a(u^1,u^2)$; (4) the set of all parameterized surfaces $x^a = x^a(u^1,u^2)$, $(u^1,u^2) \in D$, of class c^H which lie entirely in R and terminate in the given curve C_0. (Here H is the functional order of the expression

$$\sum_{\alpha} (-1)^{\alpha_1}(-1)^{\alpha_2} E_{(\alpha_1,\alpha_2)a}{}^{(\alpha_1,\alpha_2)}\ .)$$

Determine an admissible surface $\mathcal{S}_0$: $x^a = x_0^a(u^1,u^2)$ such that if S_v, given by $\mathcal{S}_v$: $x^a = x^a(u^1,u^2,v)$ is a set of admissible surfaces with

$x^a(u^1,u^2,v)\Big|_{\text{on } C} = x_0^a(u^1,u^2)$, then for $E_{\alpha a}$ evaluated along $\mathcal{S}_0$,

$$(13.8)\qquad \iint_D \sum_\alpha E_{(\alpha_1,\alpha_2)a} V^{a(\alpha_1,\alpha_2)} du^1 du^2 = 0$$

where the V^a are variation functions associated with the set S_v of surfaces.

It develops that the necessary and sufficient condition that $\mathcal{S}_0$ be such a 'normalizing surface' is formally identical to (13.3), and that in case $E_{(\alpha_1,\alpha_2)a}$ is the gradient extensor $F_{;(\alpha_1,\alpha_2)a}$ for the integrand of the ordinary double integral

$$(13.9)\qquad I = \iint_{D_0} F(x,x^{(0,1)},x^{(1,0)}, \ldots ,x^{(0,M)}, \ldots ,x^{(M,0)})du^1 du^2,$$

then the tensors $S_a^{[0,1]}(F_{;\alpha a})$, $S_a^{[1,0]}(F_{;\alpha a})$, . . . , $S_a^{[0,M]}(F_{;\alpha a})$, . . . , $S_a^{[M,0]}(F_{;\alpha a})$ appear in the condition (13.10) that I shall be invariant with respect to admissible transformations of parameter pairs:

$$(13.10)\qquad F = \sum_\gamma \gamma_2 \left[S_a^{[\gamma_1,\gamma_2]}(F_{;\alpha a})x^{(0,1)a}\right]^{(\gamma_1,\gamma_2-1)}$$
$$+ \sum_\gamma \gamma_1 \left[S_a^{[\gamma_1,\gamma_2]}(F_{;\alpha a})x^{(0,1)a}\right]^{(\gamma_1-1,\gamma_2)} .$$

Within (13.10) the summation on γ in each term is performed over all (γ_1,γ_2) such that $\gamma_1 \geq 0$, $\gamma_2 \geq 0$, $\gamma_1 + \gamma_2 \leq M$ excluding $\gamma_1 = \gamma_2 = 0$.

Here, for each (γ_1,γ_2) such that $\gamma_1 + \gamma_2 > 0$ and $\gamma_1 + \gamma_2 \leq M$, the tensor $S_a^{[\gamma_1,\gamma_2]}(F_{;\alpha a})$ is given by

(13.11)

$$S_a^{[\gamma_1,\gamma_2]}(F_{;\alpha a}) = \sum_{\alpha} (-1)^{\alpha_1}(-1)^{\alpha_2}\binom{\alpha_1+\gamma_1}{\gamma_1}\binom{\alpha_2+\gamma_2}{\gamma_2}F_{;(\alpha_1+\gamma_1,\alpha_2+\gamma_2)a}^{(\alpha_1,\alpha_2)}$$

where the α-summation is performed over all (α_1,α_2) such that $\alpha_1 + \gamma_1 \leq M$, $\alpha_2 + \gamma_2 \leq M$, and $\alpha_1 + \alpha_2 + \gamma_1 + \gamma_2 \leq M$.

Therefore, we encounter the Hamiltonian tensor $S_a^{[0]}(E_{\alpha a})$ (or $S_a^{[0,0]}(E_{\alpha a})$) in generalized simple problems which involve the variational integral $\int_{t_1}^{t_2} \sum_{\alpha=0}^{M} E_{\alpha a}V^{a(\alpha)}dt \left(\text{or } \iint_{D_0} \sum_{\alpha} E_{\alpha a}V^{a(\alpha)}du^1du^2\right)$. Further, we come upon a set S_s of tensors associated with $E_{\alpha a}$ in ordinary simple problems of the type $\int_{t_1}^{t_2} Fdt$ (or $\iint_{D_0} Fdu^1du^2$) merely by choosing the gradient extensor $F_{;\alpha a}$ as generative extensor $E_{\alpha a}$. Here, the set S_s of tensors $S_a^{[0]}(F_{;\alpha a})$, $S_a^{[1]}(F_{;\alpha a})$, . . . , $S_a^{[M]}(F_{;\alpha a})$ becomes precisely the set of Synge tensors given in [17]. (The set $S_a^{[0,0]}(F_{;\alpha a})$, $S_a^{[0,1]}(F_{;\alpha a})$, $S_a^{[1,0]}(F_{;\alpha a})$, . . . , $S_a^{[0,M]}(F_{;\alpha a})$, . . . , $S_a^{[M,0]}(F_{;\alpha a})$, of course, becomes a two-parameter analog to the set of Synge tensors given in [17]. In Chapter VI we again come upon the set S_s of tensors, but in a different variational role.

14. Extensors associated with $f(x,y_a)$ - f invariant of weight zero and y covariant--and a dynamical application. Given the function $f(x,y_a)$ with f invariant of weight zero and y covariant, the presumption of the equality of a certain pair of contravariant tensors makes possible the construction by differentiation of an excovariant extensor associated with $f(x,y_a)$. A second excovariant extensor may then be manufactured by Kawaguchi's theorem. First, f is defined by invariance, so that in coordinate system $(\bar{x})$ we have $\bar{f}$ given by

$$(14.1) \qquad \bar{f}(\bar{x},\bar{y}_r) = f\left[x(\bar{x}),\ \bar{y}_r X^r_a\right] .$$

By differentiating both members of (14.1) partially with respect to $\bar{x}^s$ with the variables $\bar{y}$ held constant, $s = 1, 2, \ldots, N$, we have

$$(14.2) \qquad \left.\frac{\partial \bar{f}}{\partial \bar{x}^s}\right|_{\bar{y}} = \left.\frac{\partial f}{\partial x^a}\right|_y \left.\frac{\partial x^a}{\partial \bar{x}^s}\right|_{\bar{y}} + \left.\frac{\partial f}{\partial y_b}\right|_x \left.\frac{\partial y_b}{\partial \bar{x}^s}\right|_{\bar{y}} .$$

We now observe that $\left.\frac{\partial y_b}{\partial \bar{x}^s}\right|_{\bar{y}}$ can be replaced according to

$$(14.3) \qquad \left.\frac{\partial y_b}{\partial \bar{x}^s}\right|_{\bar{y}} = \frac{\partial}{\partial \bar{x}^s}\left(\bar{y}_r X^r_b\right) = \bar{y}_r \frac{\partial}{\partial \bar{x}^s}\left(X^r_b\right)$$

$$= \bar{y}_r X^r_{ba} X^a_s .$$

Also, we define the symbols $f^{;b}$ according to

$$(14.4) \qquad f^{;b} \overset{\text{def}}{\equiv} \left.\frac{\partial f}{\partial y_b}\right|_x \quad .$$

This choice of notation in (14.4) is consistent with the contravariance of the quantities $\left.\frac{\partial f}{\partial y_b}\right|_x$, i.e.,

$$(14.5) \qquad \bar{f}^{;r} = \frac{\partial \bar{f}}{\partial \bar{y}_r} = \frac{\partial f}{\partial y_b}\frac{\partial y_b}{\partial \bar{y}_r} = \sum_{b=1}^{N} \frac{\partial f}{\partial y_b} X^r_b = f^{;b} X^r_b \; .$$

Therefore, substitution of (14.3) and (14.4) into (14.2) yields

$$(14.6) \qquad - \left.\frac{\partial \bar{f}}{\partial \bar{x}^s}\right|_{\bar{y}} = - \left.\frac{\partial f}{\partial x^a}\right|_y X^a_s - f^{;b} \bar{y}_r X^r_{ba} X^a_s \quad .$$

Now the second term in the right member of (14.6) can be rewritten according to the result

$$(14.7) \qquad X^r_{ba} x'^b X^a_s = - X^r_a X^{a\prime}_s \quad ,$$

which follows from $0 = \delta^{r\prime}_s = \left(X^r_a X^a_s\right)' = X^{r\prime}_a X^a_s + X^r_a X^{a\prime}_s = X^r_{ab} x'^b X^a_s + X^r_a X^{a\prime}_s$ $= X^r_{ba} x'^b X^a_s + X^r_a X^{a\prime}_s$, provided the basic presumption of the equality of a certain pair of contravariant tensors is made according to

(14.8) $\quad f^{;b} = x'^{b}$.

Equation (14.6) now becomes

$$(14.9) \qquad -\left.\frac{\partial \bar{f}}{\partial \bar{x}^s}\right|_{\bar{y}} = -\left.\frac{\partial f}{\partial x^a}\right|_y X^a_s + \bar{y}_r X^r_a X^{a'}_s \ .$$

By the covariance of y we have $y_a = \bar{y}_r X^r_a$. Since $X^a_s = X^{0a}_{0s}$ and $X^{a'}_s = X^{1a}_{0s}$ we are permitted to write (14.9) in the form

$$(14.10) \qquad -\left.\frac{\partial \bar{f}}{\partial \bar{x}^s}\right|_{\bar{y}} = -\left.\frac{\partial f}{\partial x^a}\right|_y X^{0a}_{0s} + y_a X^{1a}_{0s} \ .$$

The quantities y_a transform according to

$$(14.11) \qquad \bar{y}_s = y_a X^a_s = y_a X^{1a}_{1s} \ .$$

By equations (14.10) and (14.11) we see that the quantities $-\left.\frac{\partial f}{\partial x^a}\right|_y$, y_a are the components of an extensor of range $\alpha : 0,1$. A second absolute excovariant extensor may be constructed therefrom by Kawaguchi's theorem. Accordingly, we now adopt

Definition (14.1). The quantities $F_{\alpha a}$ and $S_{\alpha a}$ with $\alpha : 0,1$ are called the primary extensors associated with the invariant function $f(x,y_a)$, y covariant, and a curve C along which $f^{;a} = x'^{a}$. Specifically,

$$F_{0a} = -\left.\frac{\partial f}{\partial x^a}\right|_y, \; F_{1a} = y_a; \; S_{0a} = y_a{}', \; S_{1a} = y_a \; .$$

Comment: We now proceed to construct the primary extensors associated with the function $F(x,x')$ which is the Legendre transform of $f(x,y_a)$ with $y_a = y_a(x,x')$. The Legendre transform function $F(x,x')$ is defined by

$$(14.12) \quad F(x,x') = x'^b \, y_b - f(x,y_a) \quad .$$

The primary extensors $F_{;\alpha a}$ and $R_{\alpha a}$ (with $\alpha : 0,1$) are given as usual by $F_{;\alpha a} \equiv \frac{\partial F}{\partial x^{(\alpha)a}}$ and by $R_{\alpha a} \equiv \binom{1}{A} F_{;1a}{}^{\cdot(1-A)}$, $A = \alpha$. By equation (14.12) $F_{;0a} = \left.\frac{\partial F(x,x')}{\partial x^a}\right|_{x'} = \sum_{b=1}^{N} x'^b \frac{\partial y_b}{\partial x^a} - \left.\frac{\partial f}{\partial x^a}\right|_y - \sum_{b=1}^{N} f^{;b} \frac{\partial y_b}{\partial x^a}$ $= -\left.\frac{\partial f}{\partial x^a}\right|_y$, and $F_{;1a} = \left.\frac{\partial F(x,x')}{\partial x'^a}\right|_x = y_a$, by (14.8). So, $R_{0a} = y_a{}'$, $R_{1a} = y_a$. Thus we have

Theorem (14.1). The primary extensors $F_{;\alpha a}$ and $R_{\alpha a}$ (with $\alpha : 0,1$) associated with the Legendre transform function $F(x,x')$ of $f(x,y_a)$, given by

$$F(x,x') = x'^a y_a - f(x,y_a) \quad ,$$

are equal, respectively, to the primary extensors $F_{\alpha a}$ and $S_{\alpha a}$ associated with $f(x,y_a)$.

Comment: We now proceed to an application involving the dynamical system $T(q,q')$, $V(q)$. Here, the q's are space coordinates, the potential energy function $V(q)$ depends upon the q's alone, and the kinetic energy function $T(q,q')$ is given by a quadratic form in the q' 's, namely $T = \frac{1}{2} g_{ab}q'^a q'^b$. The independent variable t is time, and the metric tensor g_{ab} is given by $g_{ab} = \rho_a \cdot \rho_b$, where ρ is the radius vector and $\rho_a = \frac{\partial\rho}{\partial q^a}$, $\rho_b = \frac{\partial\rho}{\partial q^b}$.

If we define generalized momentum coordinates p_a by $p_a \equiv T_{;1a}$ so that $p_a = g_{ab}q'^b$, it is possible to express a total energy function $h(q,p)$ in Hamiltonian form, $h(q,p) = t(q,p) + V(q)$. Because the quantities p_a may be regarded as the covariant description of the q'^a relative to g_{ab} for our mechanical system according to $p_a = g_{ab}q'^b$, $q'^a = g^{ab}p_b$, where $g^{ab} \equiv \frac{c_{ab}}{G}$, with c_{ab} the cofactor of g_{ab} and G the determinant of the g_{ab}'s, we have a kinetic energy function $t(q,p)$ replacing $T(q,q')$ <u>via</u> $T(q,q') = \frac{1}{2} g_{ab}q'^a q'^b = \frac{1}{2} g_{ab}g^{ac}p_c g^{bd}p_d$ $= \frac{1}{2} \delta^c_b p_c g^{bd} p_d = \frac{1}{2} p_b p_d g^{bd} = t(q,p)$. The space coordinates q and the momentum coordinates p are treated as independent variables in phase space and in Hamiltonian mechanics. The function $L(q,q')$ generated from $h(q,p)$ by a Legendre transformation

$$(14.13) \qquad L(q,q') = q'^a p_a - h(q,p)$$

is the Lagrangian function $L(q,q') = T(q,q') - V(q)$.

With regard to the construction of extensors from the function $h(q,p)$, we note that h is an invariant function of weight zero with p covariant. The basic presumption of the equality of a pair of contravariant tensors which is made to construct a first excovariant extensor from such a function takes the form

$$(14.14) \qquad \frac{\partial h}{\partial p_a} = q'^a \quad .$$

But $\left.\frac{\partial h(q,p)}{\partial p_a}\right|_q = \left.\frac{\partial t(q,p)}{\partial p_a}\right|_q = \left.\frac{\partial}{\partial p_a}\left(\frac{1}{2} p_b p_d g^{bd}\right)\right|_q = \frac{1}{2} g^{bd}\left[p_d \delta^d_a + p_b \delta^b_a\right]$

$= \frac{1}{2} g^{ab} p_b + \frac{1}{2} g^{ad} p_d = g^{ab} p_b = q'^a$, so that (14.14) is identically satisfied all along trajectories in dynamical space. Equation (14.14) expresses a first Hamiltonian equation associated with $h(q,p)$.

Consequently, the primary extensors $A_{\alpha a}$ and $B_{\alpha a}$ (of range $\alpha : 0,1$) associated with $h(q,p)$ are well defined and given by $A_{0a} = -\left.\frac{\partial h}{\partial q^a}\right|_p$, $A_{1a} = p_a$; $B_{0a} = p_a{}'$, $B_{1a} = p_a$ (see [1], p. 161, [11], p. 199). We observe that equality of the primary extensors associated with $h(q,p)$, i.e., $A_{\alpha a} = B_{\alpha a}$, expresses a second Hamiltonian equation of motion, namely

$$(14.15) \qquad -\left.\frac{\partial h}{\partial q^a}\right|_p = p_a{}' \ .$$

Now the Lagrangian function $L(q,q')$ given by (14.13) has the properties $\left.\frac{\partial L}{\partial q^a}\right|_{q'} = -\left.\frac{\partial h}{\partial q^a}\right|_p$, $\left.\frac{\partial L}{\partial q'^a}\right|_p = p_a$. If we define $L_{;0a}$ and $L_{;1a}$ by $L_{;0a} = \left.\frac{\partial L}{\partial q^a}\right|_{q'}$ and $L_{;1a} = \left.\frac{\partial L}{\partial q'^a}\right|_q$, respectively, then it follows that the primary extensors (of range $\alpha : 0,1$) $L_{;\alpha a}$ and $M_{\alpha a}\left(\equiv \binom{1}{A} L_{;1a}{}^{\cdot(1-A)},\ A = \alpha\right)$ associated with $L(q,q')$ are equal, respectively, to the primary extensors associated with $h(q,p)$. Accordingly, the second Hamiltonian equation (14.15) can be expressed as the equality of the primary extensors associated with the Lagrangian function $L(q,q') = T(q,q') - V(q)$. Thus an equation of motion in Lagrangian form is obtainable as the set of Euler equations associated with the ordinary simple calculus of variations problem of the type $\int_{t_1}^{t_2} L(q,q')dt$.

A somewhat different viewpoint which leads to the equations of motion in Lagrangian form as in the above is to assume a linear relationship exists among the primary extensors associated with the two key functions $T(q,q')$ and $V(q)$ for our conservative dynamical system, thus

$$(14.16) \qquad A\, T_{;\alpha a} + T_{\alpha a} + B\, V_{;\alpha a} = 0$$

where A and B are constants, and where the operator $;\alpha a$ denotes $\frac{\partial}{\partial q^{(\alpha)a}}$. Our rationale for this assumption is our intuitive feeling that the differential equations of motion should be expressible in a simple manner, in terms of rates of change of the basic functions, in a form independent of the coordinate system. (Of course, such a form is the Hamiltonian form, already noted.)

The primary extensors $T_{;\alpha a}$, $T_{\alpha a}$, and $V_{;\alpha a}$ (of range α : 0,1) are given by ranks as follows:

		$\alpha=0$	$\alpha=1$
$T_{;\alpha a}$	:	$T_{;0a}$	$T_{;1a}$
$T_{\alpha a}$	:	$T_{;1a}{}'$	$T_{;1a}$
$V_{;\alpha a}$	:	$V_{;0a}$	0

By rank $\alpha = 1$ of (14.16) there follows

$$(14.17) \qquad A\,T_{;1a} + T_{;1a} = 0 \quad .$$

We see that we may avoid an imposition of excessive conditions on the functions T and V by regarding (14.17) as an identity in q and q', implying $A = -1$. We now proceed to determine B. The extensor equation (14.16) becomes

$$(14.18) \qquad -\,T_{;\alpha a} + T_{\alpha a} + B\,V_{;\alpha a} = 0 \quad .$$

Now for each $a(= 1, 2, \ldots, N)$ we multiply (14.18) by q'^a. We then sum the resulting N equations, obtaining

$$(14.19) \qquad -\,q'^a T_{;0a} + q'^a T_{;1a}{}' + Bq'^a V_{;0a} = 0 \quad .$$

Let us now impose the simple physical assumption of the conservation of energy, i.e., $T' + V' = 0$, which is equivalent to

$$(14.20) \qquad T_{;0a}q'^a + T_{;1a}q''^a + V_{;0a}q'^a = 0 \quad .$$

Now by $T = \frac{1}{2} g_{ab}q'^a q'^b$ we have that $T(q,q')$ is a homogeneous function of degree two in the variables q', so that we have

$$(14.21) \qquad q'^a \, T_{;1a} = 2T \quad .$$

By differentiation of (14.21) with respect to the parameter t we get

$$(14.22) \qquad T_{;1a}{}' \, q'^a + T_{;1a}q''^a = 2T' \quad .$$

Combining (14.19), (14.20), and (14.22) we now have

$$(14.23) \qquad 2T' + (B+1)q'^a V_{;0a} = 0 \quad .$$

By $V' = V_{;0a}q'^a$ this becomes

$$(14.24) \qquad 2T' + (B+1)V' = 0 \quad ,$$

and by the conservation equation $T' + V' = 0$ we have from (14.24) the result $B = 1$. The linear extensor equation to be examined (14.18) now becomes

$$(14.25) \quad - T_{;\alpha a} + T_{\alpha a} + V_{;\alpha a} = 0 \ .$$

The only nonvanishing rank of (14.25) is rank $\alpha = 0$, which expresses the tensor equations

$$(14.26) \quad - T_{;0a} + T_{;1a}{}' + V_{;0a} = 0 \ .$$

These Lagrangian equations of motion are equivalent to (14.15), but have been obtained on the basis of the assumption of a linear relationship among the primary extensors associated with T and V as basic functions.

The extensor approach admits of other procedures for obtaining the equations of motion for a conservative dynamical system, one of which is to assume a linear relationship among the primary extensors associated with the two key functions $\left[T(q,q')\right]^{(Q)}$ and $\left[V(q)\right]^{(Q)}$ where Q is an arbitrary positive integer. After the determination of a constant by the conservation of energy condition, this amounts to equating the primary extensors associated with $\left[L(q,q')\right]^{(Q)}$, where Q is an arbitrary positive integer and where $L(q,q') = T(q,q') - V(q)$. The fact that Lagrangian equations of motion result from this method should be contrasted to the fact that the Euler equations associated with the ordinary simple calculus of variations problem of the type $\int_{t_1}^{t_2} \left[L(q,q')\right]^{(Q)} dt$ vanish identically.

By Gammel's theorem [21] the gradient extensor $L^{(Q)}{}_{;\alpha a}$ associated with the function $L^{(Q)}$ is given by

$$(14.27) \qquad L^{(Q)}{}_{;\alpha a} = \binom{Q}{\alpha} L_{;0a}{}^{(Q-\alpha)} + \binom{Q}{\alpha-1} L_{;1a}{}^{(Q-\alpha+1)}$$

for $\alpha = 0, 1, \ldots, Q, Q+1$. The tensor rank of $L^{(Q)}{}_{;\alpha a}$ is rank $\alpha = Q + 1$ and is given by $L^{(Q)}{}_{;(Q+1)a} = L_{;1a}$. Accordingly, the Kawaguchi extensor $L^{(Q)}{}_{\alpha a}$ associated with the function $L^{(Q)}$ is given by

$$(14.28) \qquad L^{(Q)}{}_{\alpha a} = \binom{Q+1}{A} L_{;1a}{}^{\cdot(Q+1-A)}, \qquad A = \alpha,\ \alpha = 0, 1, \ldots, Q, Q+1.$$

We then have by (14.27) and (14.28) that the equality of the primary extensors associated with $\left[L(q,q')\right]^{(Q)}$, Q an arbitrary positive integer, gives us

$$(14.29) \qquad \binom{Q}{\alpha} L_{;0a}{}^{(Q-\alpha)} = \left[-\binom{Q}{\alpha-1} + \binom{Q+1}{\alpha}\right] L_{;1a}{}^{\cdot(Q+1-\alpha)}$$
$$= \binom{Q}{\alpha} L_{;1a}{}^{\cdot(Q+1-\alpha)} .$$

Equation (14.29) can be written in the form

$$(14.30) \qquad \binom{Q}{\alpha}\left[L_{;0a} - L_{;1a}{}'\right]^{(Q-\alpha)} = 0$$

for $\alpha = 0, 1, \ldots, Q$. The familiar Lagrangian equations of motion $L_{;0a} - L_{;1a}{}' = 0$, and their derivatives, are the result of (14.30).

CHAPTER II

EXTENSORS AND THE HIGHER VARIATIONS OF CERTAIN CALCULUS OF VARIATIONS INTEGRALS

15. <u>A two-parameter extensor-based Kawaguchi theorem.</u> It will be our purpose in this chapter to express the integrands of the higher variations of certain calculus of variations integrals as contractions of two-parameter or three-parameter matrix extensors. The following theorem, which may be regarded as a two-parameter extensor-based Kawaguchi theorem, will be used in obtaining these expressions.

Theorem (15.1). Let there be given the class C^Q two-parameter extensor $T_{\alpha\beta\cdot a}$ of range $0 \leq \alpha \leq M$, $\beta = 0$, defined along a class C^{M+Q} surface, so that for each admissible complete class C^{M+Q} coordinate transformation from a specified x-system in our given N-dimensional space to a second x-system we have

$$T_{\alpha\beta\cdot a} = T_{\alpha 0\cdot a} = \sum_{\rho=0}^{M} \sum_{\sigma=0}^{0} \overline{T}_{\rho\sigma\cdot r} X^{\rho\sigma\cdot r}_{\alpha 0\cdot a} = \sum_{\rho=0}^{M} \overline{T}_{\rho 0\cdot r} X^{\rho 0\cdot r}_{\alpha 0\cdot a} .$$

Then the quantities $E_{\alpha\beta\cdot a}$, defined for $0 \leq \alpha \leq M$, $0 \leq \beta \leq Q$, by

$$E_{\alpha\beta\cdot a} \overset{\text{def}}{\equiv} \binom{Q}{\beta} T_{\alpha 0\cdot a}^{(0,Q-\beta)}$$

are the components of a two-parameter extensor of range $0 \leq \alpha \leq M$, $0 \leq \beta \leq Q$.

Proof: From the definition of the quantities $E_{\alpha\beta\cdot a}$ and from the given two-parameter extensor structure of the symbols $T_{\alpha 0\cdot a}$ we have

$$E_{\alpha\beta\cdot a} = \binom{Q}{\beta} \left[\sum_{\rho=0}^{M} \overline{T}_{\rho 0\cdot r} X_{\alpha 0\cdot a}^{\rho 0\cdot r} \right]^{(0,Q-\beta)}$$

for $0 \leq \alpha \leq M$, $0 \leq \beta \leq Q$. By the Leibnitz differentiation formula this becomes

$$E_{\alpha\beta\cdot a} = \binom{Q}{\beta} \sum_{\rho=0}^{M} \sum_{\mu=0}^{Q-\beta} \binom{Q-\beta}{\mu} \overline{T}_{\rho 0\cdot r}^{(0,Q-\beta-\mu)} X_{\alpha 0\cdot a}^{\rho 0\cdot r(0,\mu)} .$$

By the substitution $X_{\alpha 0\cdot a}^{\rho 0\cdot r(0,\mu)} = \binom{\mu+\beta}{\beta}^{-1} X_{(\alpha,\beta)a}^{(\rho,\mu+\beta)r}$ we next obtain

$$E_{\alpha\beta\cdot a} = \binom{Q}{\beta} \sum_{\rho=0}^{M} \sum_{\mu=0}^{Q-\beta} \binom{Q-\beta}{\mu} \overline{T}_{\rho 0\cdot r}^{(0,Q-\beta-\mu)} \binom{\mu+\beta}{\beta}^{-1} X_{(\alpha,\beta)a}^{(\rho,\mu+\beta)r} .$$

We now replace the index μ of summation with the index σ according to $\mu = \sigma - \beta$. We have

$$E_{\alpha\beta\cdot a} = \binom{Q}{\beta} \sum_{\rho=0}^{M} \sum_{\sigma=\beta}^{Q} \binom{Q-\beta}{\sigma-\beta} \binom{\sigma}{\beta}^{-1} \overline{T}_{\rho 0\cdot r}{}^{(0,Q-\sigma)} X^{\rho\sigma\cdot r}_{\alpha\beta\cdot a} \quad .$$

By using $\binom{Q}{\beta}\binom{Q-\beta}{\sigma-\beta}\binom{\sigma}{\beta}^{-1} = \binom{Q}{\sigma}$ we may write

$$E_{\alpha\beta\cdot a} = \sum_{\rho=0}^{M} \sum_{\sigma=\beta}^{Q} \binom{Q}{\sigma} \overline{T}_{\rho 0\cdot r}{}^{(0,Q-\sigma)} X^{\rho\sigma\cdot r}_{\alpha\beta\cdot a} \quad .$$

Now the property $X^{\rho\sigma\cdot r}_{\alpha\beta\cdot a} = 0$ whenever $\sigma < \beta$ permits us to commence the σ-summation at the lower limit 0, instead of at β, $\beta \geq 0$, as the additional terms that are introduced will vanish. Further, because by definition of the symbols E there results $\overline{E}_{\rho\sigma\cdot r} = \binom{Q}{\sigma}\overline{T}_{\rho 0\cdot r}{}^{(0,Q-\sigma)}$ we have that

$$E_{\alpha\beta\cdot a} = \sum_{\rho=0}^{M} \sum_{\sigma=0}^{Q} \overline{E}_{\rho\sigma\cdot r} X^{\rho\sigma\cdot r}_{\alpha\beta\cdot a}$$

for $0 \leq \alpha \leq M$, $0 \leq \beta \leq Q$. Thus the quantities $E_{\alpha\beta\cdot a}$ follow the transformation rule of a two-parameter matrix-extensor of range $0 \leq \alpha \leq M$, $0 \leq \beta \leq Q$, and the theorem is proved.

16. The second variation of $\int_{t_1}^{t_2} F(x,x', \ldots, x^{(M)})dt$.

In this section the ordinary simple calculus of variations problem involving $I = \int_{t_1}^{t_2} F(x, x', \ldots, x^{(M)})dt$ will be formulated briefly.

In the context of this formulation the first variation of I is introduced. The higher variations of I are then defined in an obvious manner. Thus we assume there is given for consideration the following calculus of variations problem: Given (1) a function $F(x, x', \ldots, x^{(M)})$ of class C^{M+1} for x in a connected region R, (2) the interval $t_1 \leq t \leq t_2$ on an auxiliary t-axis, (3) a pair of fixed points P and Q in R with coordinates x_1^a, x_2^a, and (4) the set $\mathcal{C}$ of arcs $C : x^a = x^a(t)$, $t_1 \leq t \leq t_2$, which are of class C^{2M}, lie entirely in R, and satisfy both the requirement $\sum_{a=1}^{N} (x'^a)^2 > 0$ and the boundary conditions $x^{(\alpha)}(t_1) = x_1^{\alpha a}$, $x^{(\alpha)}(t_2) = x_2^{\alpha a}$ for $0 \leq \alpha \leq M - 1$. Find an arc C_0 in the set $\mathcal{C}$ of arcs such that C_0 furnishes a stationary value to the integral $\int_{t_1}^{t_2} F(x, x', \ldots, x^{(M)})dt$.

With regard to the solution of this problem, we have for $\{C_v : x^a = x^a(t,v)\}$ a set of arcs in $\mathcal{C}$, for $V^a(t)$ a set of variation functions such that $V^{a(\Omega)}(t_1) = V^{a(\Omega)}(t_2) = 0$ for $0 \leq \Omega \leq M - 1$, and for v a variation parameter, the result that it is possible (see [19]) to express the first variation of

$$(16.1)\qquad I(v) = \int_{t_1}^{t_2} F\left[x^a(t,v), x^{a'}(t,v), \ . \ . \ . \ , \ x^{a(M)}(t,v)\right] dt$$

as the integral of a contraction of one-parameter extensors with parameter t. Specifically, we have

$$(16.2)\qquad \frac{dI(v)}{dv} = \int_{t_1}^{t_2} \sum_{\alpha=0}^{M} F_{;\alpha a} V^{a(\alpha)} dt \quad .$$

Nevertheless, if $V^a \equiv \frac{\partial x^a}{\partial v}$, then $V^a = V^a(t,v)$, so that the integrand function of (16.2) depends upon the variation parameter v as well as upon the arc parameter t. By (16.2) and by the continuity of its integrand with respect to v we may form the second variation $\frac{d^2I(v)}{dv^2}$ of (16.1) according to

$$(16.3)\qquad \frac{d^2I(v)}{dv^2} = \int_{t_1}^{t_2} \frac{\partial}{\partial v}\left[\sum_{\alpha=0}^{M} F_{;\alpha a} \frac{\partial x^{(\alpha)a}}{\partial v}\right] dt$$

where the integrand of (16.3) is evaluated along curve number v in the set $\{C_v\}$. By performing the indicated differentiation in (16.3) we get

$$(16.4)\qquad \frac{d^2I(v)}{dv^2} = \int_{t_1}^{t_2}\left[\sum_{\alpha=0}^{M}\sum_{\beta=0}^{M} F_{;\alpha a;\beta b} \frac{\partial x^{a(\alpha)}}{\partial v} \frac{\partial x^{b(\beta)}}{\partial v}\right.$$

$$\left. + \sum_{\alpha=0}^{M} F_{;\alpha a} \frac{\partial^2 x^{a(\alpha)}}{\partial v^2}\right] dt \quad .$$

In two-parameter notation, with the first parameter t and the second parameter v, we may write

$$(16.5)\qquad \frac{d^2 I(v)}{dv^2} = \int_{t_1}^{t_2} \sum_{\alpha_1=0}^{M} \sum_{\alpha_2=0}^{1} E_{(\alpha_1,\alpha_2)a} V^{a(\alpha_1,\alpha_2)} dt$$

where $E_{(\alpha_1,\alpha_2)a}$ is defined by

$$(16.6)\qquad E_{(\alpha_1,\alpha_2)a} = \binom{1}{\alpha_2} F_{;(\alpha_1,0)a}^{(0,1-\alpha_2)}$$

for $0 \leq \alpha_1 \leq M$, $0 \leq \alpha_2 \leq 1$, and where $V^{a(\alpha_1,\alpha_2)} = x^{a(\alpha_1,\alpha_2+1)}$ has been utilized.

Now, by construction, $V^{a(\alpha_1,\alpha_2)}$ is a two-parameter excontravariant extensor of range $0 \leq \alpha_1 \leq M$, $0 \leq \alpha_2 \leq 1$. Further, by Theorem (15.1), the quantities $E_{(\alpha_1,\alpha_2)a}$ of equation (16.6) are the components of a two-parameter excovariant extensor of range $0 \leq \alpha_1 \leq M$, $0 \leq \alpha_2 \leq 1$. The integrand of (16.5), which is an invariant by construction, is, therefore, a full-range extensor contraction over the range $0 \leq \alpha_1 \leq M$, $0 \leq \alpha_2 \leq 1$. Accordingly, (16.5) expresses the second variation of (16.1) as the integral of a contraction of two-parameter extensors.

17. The higher variations of $\int_{t_1}^{t_2} F(x, x', \ldots, x^{(M)})dt$.

With regard to the ordinary simple calculus of variations problem of Section 16 with $V^a = \frac{\partial x^a}{\partial v}$, $V^a = V^a(t,v)$, and v a variation parameter, it is possible to form from (16.2) an expression for the (K+1)st variation of I, $K \geq 0$, thus:

$$(17.1) \qquad \frac{d^{K+1}I(v)}{dv^{K+1}} = \int_{t_1}^{t_2} \frac{\partial^K}{\partial v^K} \sum_{\alpha=0}^{M} \left[F_{;\alpha a} \frac{\partial x^{(\alpha)a}}{\partial v} \right] dt \; .$$

Here the integral I(v) is given by (16.1), and, subject only to differentiability limitations, K may assume any non-negative integral value. The integrand of (17.1) is evaluated along curve number v in the set $\{C_v : x^a = x^a(t,v)\}$ of arcs.

In two-parameter notation, with the first parameter t and the second parameter v, we may write (17.1) in the form

$$(17.2) \qquad \frac{d^{K+1}I(v)}{dv^{K+1}} = \int_{t_1}^{t_2} \left[\sum_{\alpha_1=0}^{M} F_{;(\alpha_1,0)a} x^{(\alpha_1,1)a} \right]^{(0,K)} dt.$$

The differentiation K times with respect to v in the integrand of the right member of (17.2) may be performed by the Leibnitz formula, yielding

$$(17.3)\quad \frac{d^{K+1}I(v)}{dv^{K+1}} = \int_{t_1}^{t_2} \sum_{\alpha_1=0}^{M} \sum_{\mu=0}^{K} \binom{K}{\mu} F^{(0,\mu)}_{;(\alpha_1,0)a} \left[x^{a(\alpha_1,1)}\right]^{(0,K-\mu)} dt.$$

Noting that $\left[x^{a(\alpha_1,1)}\right]^{(0,K-\mu)} = x^{a(\alpha_1,K-\mu+1)}$ and that $x^{a(\alpha_1,K-\mu+1)} = V^{a(\alpha_1,K-\mu)}$, we may obtain from (17.3) the result

$$(17.4)\quad \frac{d^{K+1}I(v)}{dv^{K+1}} = \int_{t_1}^{t_2} \sum_{\alpha_1=0}^{M} \sum_{\mu=0}^{K} \binom{K}{\mu} F^{(0,\mu)}_{;(\alpha_1,0)a} V^{a(\alpha_1,K-\mu)} dt .$$

Replacing the summation index μ in the right member of (17.4) by α_2 according to $K - \mu = \alpha_2$, we get

$$(17.5)\quad \frac{d^{K+1}I(v)}{dv^{K+1}} = \int_{t_1}^{t_2} \left[\sum_{\alpha_1=0}^{M} \sum_{\alpha_2=0}^{K} \binom{K}{\alpha_2} F^{(0,K-\alpha_2)}_{;(\alpha_1,0)a}\right] V^{a(\alpha_1,\alpha_2)} dt.$$

Accordingly, we may write

$$(17.6)\quad \frac{d^{K+1}I(v)}{dv^{K+1}} = \int_{t_1}^{t_2} \sum_{\alpha_1=0}^{M} \sum_{\alpha_2=0}^{K} E_{(\alpha_1,\alpha_2)a} V^{a(\alpha_1,\alpha_2)} dt$$

where $E_{(\alpha_1,\alpha_2)a}$ is defined by

$$(17.7)\quad E_{(\alpha_1,\alpha_2)a} \overset{\text{def}}{\equiv} \binom{K}{\alpha_2} F^{(0,K-\alpha_2)}_{;(\alpha_1,0)a} .$$

Now, by construction, $V^{a(\alpha_1,\alpha_2)}$ is a two-parameter excontravariant extensor of range $0 \leq \alpha_1 \leq M$, $0 \leq \alpha_2 \leq K$. By Theorem (15.1) the quantities $E_{(\alpha_1,\alpha_2)a}$ are the components of a two-parameter exco-variant extensor of range $0 \leq \alpha_1 \leq M$, $0 \leq \alpha_2 \leq K$. Therefore, the integrand of (17.6), which is an invariant by construction, is a full-range extensor contraction over the range $0 \leq \alpha_1 \leq M$, $0 \leq \alpha_2 \leq K$. Accordingly, (17.6) expresses the (K+1)st variation of (16.1) as the integral of a contraction of two-parameter extensors.

18. A recursion formula for the higher variations of $\int_{t_1}^{t_2} F(x, x', \ldots, x^{(M)})dt$ involving two-parameter gradient extensors.

It is interesting to note here that, if G_K is the integrand function of the Kth variation of I, I given by (16.1), then the two-parameter gradient extensor $G_{K;(\alpha_1,\alpha_2)a}$ associated with G_K appears contracted with $V^{a(\alpha_1,\alpha_2)}$ as the integrand function G_{K+1} of the (K+1)st variation of I. In other words, if we let K be a positive integer and if we define G_K and G_{K+1} such that

$$(18.1) \qquad \frac{d^K I(v)}{dv^K} = \int_{t_1}^{t_2} G_K dt$$

and

$$(18.2)\qquad \frac{d^{K+1}I(v)}{dv^{K+1}} = \int_{t_1}^{t_2} G_{K+1}\,dt$$

then there follows

$$(18.3)\qquad G_{K+1} = \sum_{\alpha_1=0}^{M} \sum_{\alpha_2=0}^{K} G_{K;(\alpha_1,\alpha_2)a} V^{a(\alpha_1,\alpha_2)} \quad .$$

This results in the following manner: First, use the definition of $\frac{d^{K+1}I(v)}{dv^{K+1}}$ to write

$$(18.4)\qquad \frac{d^{K+1}I(v)}{dv^{K+1}} = \frac{d}{dv}\left[\frac{d^{K}I(v)}{dv^{K}}\right] = \frac{d}{dv}\int_{t_1}^{t_2} G_K\,dt \quad .$$

Next, note by differentiability of G_K with respect to v that we have

$$(18.5)\qquad \frac{d^{K+1}I(v)}{dv^{K+1}} = \int_{t_1}^{t_2} G_K^{(0,1)}dt \quad .$$

Subsequently, utilize the chain rule of differentiation to replace $G_K^{(0,1)}$ by $\sum_{\alpha_1=0}^{M} \sum_{\alpha_2=0}^{K} G_{K;(\alpha_1,\alpha_2)a} x^{a(\alpha_1,\alpha_2+1)}$, obtaining

$$(18.6)\qquad \frac{d^{K+1}I(v)}{dv^{K+1}} = \int_{t_1}^{t_2} \sum_{\alpha_1=0}^{M} \sum_{\alpha_2=0}^{K} G_{K;(\alpha_1,\alpha_2)a} x^{a(\alpha_1,\alpha_2+1)}dt \quad .$$

Finally, recall that $x^{a(\alpha_1,\alpha_2+1)} = V^{a(\alpha_1,\alpha_2)}$ for $0 \le \alpha_1 \le M$, $0 \le \alpha_2 \le K+1$,

and express (18.6) in the form

$$(18.7)\qquad \frac{d^{K+1}I(v)}{dv^{K+1}} = \int_{t_1}^{t_2} \sum_{\alpha_1=0}^{M} \sum_{\alpha_2=0}^{K} G_{K;(\alpha_1,\alpha_2)a} V^{a(\alpha_1,\alpha_2)} dt.$$

Equation (18.3) follows by definition of G_{K+1}, see (18.1).

We return to consider equations (18.1), (18.2), and (18.3). These equations may be regarded as defining a recursion formula, involving two-parameter gradient extensors, for generating the successive higher variations of $I = \int_{t_1}^{t_2} F(x, x', \ldots, x^{(M)})dt$, while preserving their explicit expression in terms of an extensor structure.

19. Three-parameter extensors and the higher variations of the simple double integral $\iint_{D_0} F(x,x^{(0,1)},x^{(1,0)},\ldots,x^{(0,M)},\ldots,x^{(M,0)})du^1du^2$.

In this section an ordinary simple double integral calculus of variations problem involving $J = \iint_{D_0} F(x,x^{(0,1)},x^{(1,0)}, \ldots, x^{(0,M)}, \ldots, x^{(M,0)})du^1du^2$ is formulated briefly. The first, second, and higher variations of J are defined in the context of the problem formulation. Thus we assume there is given for consideration the following simple double integral calculus of variations problem: Given (1) a three-dimensional space X_3 bearing a coordinate system (x), (2) a function $F(x^{(\alpha_1,\alpha_2)})$, $0 \leq \alpha_1 + \alpha_2 \leq M$, of class C^{M+1}

for x in a connected region R, (3) a given closed space curve C_0 lying completely in R, (4) a domain D in an auxiliary Euclidean u^1,u^2-plane with boundary curve C, such that C_0 is the map of C by $x^a = x^a(u^1,u^2)$, and (5) the set S of surfaces $\mathcal{S}$: $x^a = x^a(u^1,u^2)$, $(u^1,u^2) \in D$, a = 1, 2, 3 (alternately $\rho = \rho(u^1,u^2)$) which are of class c^{2M} in R, have all their elements in R, and satisfy both the requirement $\frac{\partial\rho}{\partial u^1} \times \frac{\partial\rho}{\partial u^2} \neq 0$ for each (u^1,u^2) in D and the boundary condition that they terminate in C_0. Find a surface $\mathcal{S}_0$ in the set S of surfaces such that $\mathcal{S}_0$ furnishes a stationary value to the integral $\iint_{D_0} F(x, x^{(0,1)}, x^{(1,0)}, \ldots, x^{(0,M)}, \ldots, x^{(M,0)}) du^1 du^2$.

With regard to the solution of this problem, we have for $\{\mathcal{S}_v : x^a = x^a(u^1,u^2,v)\}$ a set of surfaces in S, for $V^a(u^1,u^2)$ a set of variation functions such that $V^{a(\Omega)}(u^1,u^2)\Big|_C = 0$ for $\Omega = (\Omega_1,\Omega_2)$ and $0 \leq \Omega_1 + \Omega_2 \leq M - 1$, and for v a variation parameter, the result that it is possible to express the first variation of

$$(19.1) \qquad J(v) = \iint_D F\left[x^a(u^1,u^2,v), \ldots, x^{a(M,0)}(u^1,u^2,v), \ldots, x^{a(0,M)}(u^1,u^2,v)\right] du^1 du^2$$

as the integral of a contraction of two-parameter extensors, with the two parameters u^1 and u^2, respectively (see [19]). Specifically, we have

$$(19.2)\quad \frac{dJ(v)}{dv} = \iint_D \sum_{\alpha_1+\alpha_2\geq 0}^{\alpha_1+\alpha_2\leq M} F_{;(\alpha_1,\alpha_2)a} V^{a(\alpha_1,\alpha_2)} du^1 du^2$$

where D is the domain of integration in the u^1,u^2-plane for the surface $\mathcal{S}$.

Nevertheless, if $V^a \equiv \frac{\partial x^a}{\partial v}$, then $V^a = V^a(u^1,u^2,v)$, so that the integrand function of (19.2) depends upon the variation parameter v as well as upon the surface parameters u^1 and u^2. By (19.2) and by the continuity of its integrand with respect to v we may form the second variation $\frac{d^2J(v)}{dv^2}$ of (19.1) according to

$$(19.3)\quad \frac{d^2J(v)}{dv^2} = \iint_{D_v} \frac{\partial}{\partial v}\left[\sum_{\alpha_1+\alpha_2\geq 0}^{\alpha_1+\alpha_2\leq M} F_{;(\alpha_1,\alpha_2)a} \frac{\partial x^{(\alpha_1,\alpha_2)a}}{\partial v}\right] du^1 du^2$$

where the integrand of (19.3) is evaluated over the domain D_v in the u^1,u^2-plane for the surface $\mathcal{S}_v$. By performing the indicated differentiation in (19.3) we get

(19.4)

$$\frac{d^2J(v)}{dv^2} = \iint_{D_v} \left\{ \sum_{\alpha_1+\alpha_2\geq 0}^{\alpha_1+\alpha_2\leq M} \sum_{\beta_1+\beta_2\geq 0}^{\beta_1+\beta_2\leq M} \left[F_{;(\alpha_1,\alpha_2)a;(\beta_1,\beta_2)b} \frac{\partial x^{a(\alpha_1,\alpha_2)}}{\partial v} \frac{\partial x^{b(\beta_1,\beta_2)}}{\partial v} \right] \right.$$

$$\left. + \sum_{\alpha_1+\alpha_2\geq 0}^{\alpha_1+\alpha_2\leq M} F_{;(\alpha_1,\alpha_2)a} \frac{\partial^2 x^{a(\alpha_1,\alpha_2)}}{\partial v} \right\} du^1 du^2 .$$

In three-parameter notation, with the three parameters u^1, u^2, and v, respectively, we may write

$$(19.5) \quad \frac{d^2J(v)}{dv^2} = \iint_{D_v} \sum_{\alpha_1+\alpha_2\geq 0}^{\alpha_1+\alpha_2\leq M} \sum_{\alpha_3=0}^{1} E_{(\alpha_1,\alpha_2,\alpha_3)a} V^{a(\alpha_1,\alpha_2,\alpha_3)} du^1 du^2$$

where $E_{(\alpha_1,\alpha_2,\alpha_3)a}$ is defined by

$$(19.6) \quad E_{(\alpha_1,\alpha_2,\alpha_3)a} \overset{\text{def}}{\equiv} \binom{1}{\alpha_3} F_{;(\alpha_1,\alpha_2,0)a}{}^{(0,0,1-\alpha_3)}$$

for $0 \leq \alpha_1 + \alpha_2 \leq M$, $0 \leq \alpha_3 \leq 1$, and where $V^{a(\alpha_1,\alpha_2,\alpha_3)} = x^{a(\alpha_1,\alpha_2,\alpha_3+1)}$ has been utilized in obtaining (19.5).

Now, by construction, $V^{a(\alpha_1,\alpha_2,\alpha_3)}$ is a three-parameter excontravariant extensor of the mixed coupled- and uncoupled-limit range $0 \leq \alpha_1 + \alpha_2 \leq M$, $0 \leq \alpha_3 \leq 1$. Further, by a three-parameter extensor-based Kawaguchi theorem similar to Theorem (15.1), the quantities

$E_{(\alpha_1,\alpha_2,\alpha_3)a}$ of equation (19.6) are the components of a three-parameter excovariant extensor of the mixed coupled- and uncoupled-limit range $0 \leq \alpha_1 + \alpha_2 \leq M$, $0 \leq \alpha_3 \leq 1$. The integrand of (19.5), which is an invariant by construction, is, therefore, a full-range extensor contraction over the range $0 \leq \alpha_1' + \alpha_2 \leq M$, $0 \leq \alpha_3 \leq 1$. Accordingly, (19.5) expresses the second variation of (19.1) as the integral of a contraction of three-parameter extensors.

With regard to the higher variations of J, we note that it is possible to form from (19.2) an expression for the (K+1)st variation, $K \geq 0$, of J, thus:

$$(19.7)\quad \frac{d^{K+1}J(v)}{dv^{K+1}} = \iint_{D_v} \frac{\partial^K}{\partial v^K}\left[\sum_{\alpha_1+\alpha_2\geq 0}^{\alpha_1+\alpha_2\leq M} F_{;(\alpha_1,\alpha_2)a} \frac{\partial x^{(\alpha_1,\alpha_2)a}}{\partial v}\right] du^1 du^2 .$$

In three-parameter notation, with the first parameter u^1, the second parameter u^2, and the third parameter v, we may write (19.7) in the form

$$(19.8)\quad \frac{d^{K+1}J(v)}{dv^{K+1}} = \iint_{D_v} \left[\sum_{\alpha_1+\alpha_2\geq 0}^{\alpha_1+\alpha_2\leq M} F_{;(\alpha_1,\alpha_2,0)a}\, x^{(\alpha_1,\alpha_2,1)a}\right]^{(0,0,K)} du^1 du^2 .$$

The operation superior (0,0,K) in the integrand of the right member of (19.8) may be performed by the Leibnitz formula, yielding

(19.9)

$$\frac{d^{K+1}J(v)}{dv^{K+1}} = \iint_{D_V} \sum_{\alpha_1+\alpha_2\geq 0}^{\alpha_1+\alpha_2\leq M} \sum_{\mu=0}^{K} \binom{K}{\mu} F_{;(\alpha_1,\alpha_2,0)}a^{(0,0,\mu)}\left[x^{a(\alpha_1,\alpha_2,1)}\right]^{(0,0,K-\mu)} du^1du^2.$$

Noting that $\left[x^{a(\alpha_1,\alpha_2,1)}\right]^{(0,0,K-\mu)} = x^{a(\alpha_1,\alpha_2,K-\mu+1)}$ and that $x^{a(\alpha_1,\alpha_2,K-\mu+1)} = V^{a(\alpha_1,\alpha_2,K-\mu)}$, we may obtain from (19.9) the result

(19.10)

$$\frac{d^{K+1}J(v)}{dv^{K+1}} = \iint_{D_V} \sum_{\alpha_1+\alpha_2\geq 0}^{\alpha_1+\alpha_2\leq M} \sum_{\mu=0}^{K} \binom{K}{\mu} F_{;(\alpha_1,\alpha_2,0)}a^{(0,0,\mu)}\, V^{a(\alpha_1,\alpha_2,K-\mu)} du^1du^2 .$$

Replacing the summation index μ in the right member of (19.10) by α_3 according to $K - \mu = \alpha_3$, we get

(19.11)

$$\frac{d^{K+1}J(v)}{dv^{K+1}} = \iint_{D_V} \left[\sum_{\alpha_1+\alpha_2\geq 0}^{\alpha_1+\alpha_2\leq M} \sum_{\alpha_3=0}^{K} \binom{K}{\alpha_3} F_{;(\alpha_1,\alpha_2,0)}a^{(0,0,K-\alpha_3)}\right] V^{a(\alpha_1,\alpha_2,\alpha_3)} du^1du^2.$$

Accordingly, we may write

$$(19.12)\quad \frac{d^{K+1}J(v)}{dv^{K+1}} = \iint_{D_V} \sum_{\alpha_1+\alpha_2\geq 0}^{\alpha_1+\alpha_2\leq M} \sum_{\alpha_3=0}^{K} E_{(\alpha_1,\alpha_2,\alpha_3)}a\, V^{a(\alpha_1,\alpha_2,\alpha_3)} du^1du^2$$

where $E_{(\alpha_1,\alpha_2,\alpha_3)a}$ is defined by

$$(19.13)\quad E_{(\alpha_1,\alpha_2,\alpha_3)a} \overset{\text{def}}{\equiv} \binom{K}{\alpha_3} F^{(0,0,K-\alpha_3)}_{;(\alpha_1,\alpha_2,0)a}$$

for $0 \leq \alpha_1 + \alpha_2 \leq M$, $0 \leq \alpha_3 \leq K$, and where $V^{a(\alpha_1,\alpha_2,\alpha_3)} = x^{a(\alpha_1,\alpha_2,\alpha_3+1)}$ has been utilized in obtaining (19.12).

Now, by construction, $V^{a(\alpha_1,\alpha_2,\alpha_3)}$ is a three-parameter excontravariant extensor of range $0 \leq \alpha_1 + \alpha_2 \leq M$, $0 \leq \alpha_3 \leq K$. Further, by a three-parameter version of Theorem (15.1), the quantities $E_{(\alpha_1,\alpha_2,\alpha_3)a}$ of equation (19.13) are the components of a three-parameter excovariant extensor of range $0 \leq \alpha_1 + \alpha_2 \leq M$, $0 \leq \alpha_3 \leq K$. The integrand of (19.12), which is an invariant by construction, is, therefore, a full-range extensor contraction over the range $0 \leq \alpha_1 + \alpha_2 \leq M$, $0 \leq \alpha_3 \leq K$. Accordingly, (19.12) expresses the (K+1)st variation of (19.1) as the integral of a contraction of three-parameter extensors.

20. A multiple-parameter extensor construction for the higher derivatives in a certain simple extensor-generalized calculus of variations problem. An extensor generalization of the simple calculus of

variations problem of the type $\int_{t_1}^{t_2} F(x, x', \ldots, x^{(M)})dt$ of Section 16 was formulated in [15]. This formulation is reproduced, as in Section 13, in a condensed form as follows: Assume there is given (1) an extensor $T_{\alpha a}$ of range α : 0 to M whose components are functions of $x, x', \ldots, x^{(P)}$ and are of class c^M for x in a connected region R and for all values of the x-primes which satisfy $\sum_{a=1}^{N} (x'^a)^2 > 0$; (2) the interval $t_1 \leq t \leq t_2$ on an auxiliary t-axis; (3) a pair of points P,Q in R with coordinates x_1^a, x_2^a; and (4) the set of all arcs $x^a = x^a(t)$, $t_1 \leq t \leq t_2$, of class c^H, where H is the functional order of $\sum_{\alpha=0}^{M} (-1)^\alpha E_{\alpha a}^{(\alpha)}$, which lie entirely in R, satisfy the requirement $\sum_{a=1}^{N} (x'^a)^2 > 0$ and the boundary conditions $x^{(\alpha)a}(t_1) = x_1^{\alpha a}, x^{(\alpha)a}(t_2) = x_2^{\alpha a}$ for $0 \leq \alpha \leq M - 1$. Find an admissible arc $C_0 : x^a = x^a(t)$ such that if $\{C_v : x^a = x^a(t,v)\}$ is a set of admissible arcs with $x^a(t,0) = x_0^a(t)$, then for $T_{\alpha a}$ evaluated along C_0 we have $\int_{t_1}^{t_2} T_{\alpha a} V^{a(\alpha)} dt = 0$. Here, given items (1) - (4), an admissible base arc which satisfies $I_1 = 0$, $I_1 \overset{\text{def}}{\equiv} \int_{t_1}^{t_2} T_{\alpha a} V^{a(\alpha)} dt$, is called a normalizing arc for the extensor $T_{\alpha a}$ under the given boundary conditions.

Thus, the concept of a normalizing arc with respect to the extensor $T_{\alpha a}$ as an arc along which I_1 vanishes replaces the concept of a stationary arc as an arc along which the first variation $\int_{t_1}^{t_2} F_{;\alpha a} V^{a(\alpha)} dt$ of $\int_{t_1}^{t_2} F(x, x', \ldots, x^{(M)}) dt$ vanishes; of course, if $T_{\alpha a}$ is a gradient extensor $F_{;\alpha a}$, then a normalizing arc is also a stationary arc (see [15], p. 315).

Accordingly, the second variation of $\int_{t_1}^{t_2} F(x,x', \ldots ,x^{(M)})dt$ would then be replaced by the first derivative with respect to the variation parameter v of the integral

$$(20.1) \qquad I_1(v) = \int_{t_1}^{t_2} \sum_{\alpha=0}^{M} T_{\alpha a} V^{a(\alpha)} dt \quad ,$$

where the integration is performed over arc number v in the set $\{C_v : x^a = x^a(t,v)\}$. In general, the Kth derivative of (20.1) with respect to v would replace the (K+1)st variation, $K \geq 0$, of $\int_{t_1}^{t_2} F(x, x', \ldots , x^{(M)})dt$.

Let us examine the Kth derivative of (20.1) with respect to v. We have

$$(20.2) \qquad \frac{d^K I_1(v)}{dv^K} = \int_{t_1}^{t_2} \frac{\partial^K}{\partial v^K} \left[\sum_{\alpha=0}^{M} T_{\alpha a} V^{a(\alpha)} \right] dt \quad .$$

By the Leibnitz formula this becomes, in two-parameter notation with the two parameters t and v, respectively,

$$(20.3)\quad \frac{d^K I_1(v)}{dv^K} = \int_{t_1}^{t_2} \sum_{\alpha=0}^{M} \sum_{\beta=0}^{K} \binom{K}{\beta} T_{(\alpha,0)a}{}^{(0,K-\beta)} v^{a(\alpha,\beta)} dt.$$

Here the one-parameter extensor $T_{\alpha a}$ is imbedded in a two-parameter range. Thus the two-parameter extensor $T_{(\alpha,\delta)a}$ of range $0 \leq \alpha \leq M$, $\delta = 0$ is defined along any arc C according to $T_{(\alpha,0)a} \overset{\text{def}}{\equiv} T_{\alpha a}$, with $T_{\alpha a}$ evaluated along C. Accordingly, we may write

$$(20.4)\quad \frac{d^K I_1(v)}{dv^K} = \int_{t_1}^{t_2} \sum_{\alpha=0}^{M} \sum_{\beta=0}^{K} E_{(\alpha,\beta)a} v^{a(\alpha,\beta)} dt$$

where the quantities $E_{(\alpha,\beta)a}$, defined by

$$(20.5)\quad E_{(\alpha,\beta)a} \overset{\text{def}}{\equiv} \binom{K}{\beta} T_{(\alpha,0)a}{}^{(0,K-\beta)}$$

for $0 \leq \alpha \leq M$, $0 \leq \beta \leq K$, are the components of a two-parameter extensor of range $0 \leq \alpha \leq M$, $0 \leq \beta \leq K$(by a two-parameter extensor-based Kawaguchi theorem, namely Theorem (15.1)). The quantities $v^{a(\alpha,\beta)}$, $0 \leq \alpha \leq M$, $0 \leq \beta \leq K$, are the components of a two-parameter excontravariant extensor of range $0 \leq \alpha \leq M$, $0 \leq \beta \leq K$ by construction, as the v^a are the components of a contravariant tensor.

Accordingly, the integrand of (20.4), which is an invariant by construction, expresses the Kth derivative, $K \geq 0$, of the $I_1(v)$ of (20.1) as an integral of a full-range contraction of the two-parameter extensors $E_{(\alpha,\beta)a}$ and $V^{a(\alpha,\beta)}$ over the range $0 \leq \alpha \leq M$, $0 \leq \beta \leq K$.

CHAPTER III

EXTENSORS ASSOCIATED WITH $f(x,x', \ldots, x^{(M)}, y_{\theta a})$-- f INVARIANT OF WEIGHT ZERO AND y EXCOVARIANT-- AND A DYNAMICAL APPLICATION

21. The primary extensors associated with $f(x,x', \ldots, x^{(M)}, y_{\theta a})$. Given the function $f(x,x', \ldots, x^{(M)}, y_{\theta a})$ with f invariant of weight zero and y an absolute excovariant extensor, the presumption of the equality of a certain pair of excontravariant extensors makes possible the construction by differentiation of an excovariant extensor, of range 0 to M+1, associated with $f(x,x', \ldots, x^{(M)}, y_{\theta a})$. A second excovariant extensor may then be manufactured by Kawaguchi's theorem. First, f is defined by invariance, so that in coordinate system $(\bar{x})$ we have $\bar{f}$ given by

(21.1)
$$\bar{f}(\bar{x},\bar{x}', \ldots, \bar{x}^{(M)}, \bar{y}_{\rho r}) = f\left[x(\bar{x}), x'(\bar{x},\bar{x}'), \ldots, x^{(M)}(\bar{x},\bar{x}', \ldots, \bar{x}^{(M)}), \bar{y}_{\rho r} X^{\rho r}_{\alpha a}\right].$$

By differentiating both members of (21.1) partially with respect to $\bar{x}^s$ with the variables $\bar{y}$ and the $\bar{x}$-primes held constant, $s = 1, 2, \ldots, N$, we have

$$(21.2)\quad \left.\frac{\partial \bar{f}}{\partial \bar{x}^s}\right|_{\bar{y}} = \sum_{\alpha=0}^{M} \left.\frac{\partial f}{\partial x^{(\alpha)a}}\right|_{y} \left.\frac{\partial x^{(\alpha)a}}{\partial \bar{x}^s}\right|_{\bar{y}} + \sum_{\beta=0}^{M} \left.\frac{\partial f}{\partial y_{\beta b}}\right|_{x} \left.\frac{\partial y_{\beta b}}{\partial \bar{x}^s}\right|_{\bar{y}} .$$

We now observe that $\left.\frac{\partial y_{\beta b}}{\partial \bar{x}^s}\right|_{\bar{y}}$ can be replaced according to

$$(21.3)\quad \left.\frac{\partial y_{\beta b}}{\partial \bar{x}^s}\right|_{\bar{y}} = \frac{\partial}{\partial \bar{x}^s}\left[\sum_{\rho=0}^{M} \bar{y}_{\rho r} X^{\rho r}_{\beta b}\right]$$

$$= \sum_{\rho=0}^{M} \bar{y}_{\rho r} \frac{\partial}{\partial \bar{x}^s} X^{\rho r}_{\beta b}$$

$$= \sum_{\rho=0}^{M} \bar{y}_{\rho r} \sum_{\alpha=0}^{M} X^{\rho r}_{\beta b \cdot \alpha a} X^{\alpha a}_{0s} .$$

Also, we define the symbols $f^{;\beta b}$ according to

$$(21.4)\quad f^{;\beta b} \overset{\text{def}}{\equiv} \left.\frac{\partial f}{\partial y_{\beta b}}\right|_{x} .$$

This choice of notation in (21.4) is consistent with the excontravariance of the quantities $\left.\frac{\partial f}{\partial y_{\beta b}}\right|_{x}$, i.e.,

$$(21.5)\quad \bar{f}^{;\rho r} = \frac{\partial \bar{f}}{\partial \bar{y}_{\rho r}} = \sum_{\beta=0}^{M} \frac{\partial f}{\partial y_{\beta b}} \frac{\partial y_{\beta b}}{\partial \bar{y}_{\rho r}} = \sum_{\beta=0}^{M} \sum_{b=1}^{N} \frac{\partial f}{\partial y_{\beta b}} X^{\rho r}_{\beta b}$$

$$= \sum_{\beta=0}^{M} f^{;\beta b} X^{\rho r}_{\beta b} .$$

Therefore, substitution of (21.3) and (21.4) into (21.2) yields

$$(21.6)\quad -\left.\frac{\partial \bar{f}}{\partial \bar{x}^s}\right|_{\bar{y}} = -\sum_{\alpha=0}^{M} \left.\frac{\partial f}{\partial x^{(\alpha)a}}\right|_{y} X^{\alpha a}_{0s}$$

$$-\sum_{\beta=0}^{M} f^{;\beta b} \sum_{\rho=0}^{M} \bar{y}_{\rho r} \sum_{\alpha=0}^{M} X^{\rho r}_{\beta b\cdot\alpha a} X^{\alpha a}_{0s} .$$

Now the second term in the right member of (21.6) can be rewritten according to the result

$$(21.7)\quad \sum_{\beta=0}^{M}\sum_{\alpha=0}^{M} X^{\rho r}_{\beta b\cdot\alpha a} X^{\alpha a}_{0s} x'^{(\beta)b} = -\sum_{\alpha=0}^{M} X^{\rho r}_{\alpha a}\left[X^{\alpha a}_{0s}\right]',$$

which follows from $0 = \left[\delta^{\rho}_{0}\delta^{r}_{s}\right]' = \left[\sum_{\alpha=0}^{M} X^{\rho r}_{\alpha a} X^{\alpha a}_{0s}\right]' = \sum_{\alpha=0}^{M}\left[X^{\rho r}_{\alpha a}\right]' X^{\alpha a}_{0s} + \sum_{\alpha=0}^{M} X^{\rho r}_{\alpha a}\left[X^{\alpha a}_{0s}\right]'$
$= \sum_{\beta=0}^{M}\sum_{\alpha=0}^{M} X^{\rho r}_{\beta b\cdot\alpha a} X^{\alpha a}_{0s} x'^{(\beta)b} + \sum_{\alpha=0}^{M} X^{\rho r}_{\alpha a}\left[X^{\alpha a}_{0s}\right]'$, provided the basic presumption is made of the existence of an arc along which we have the equality of a certain pair of excontravariant extensors according to

$$(21.8)\quad f^{;\beta b} = x'^{(\beta)b}$$

for $\beta = 0, 1, \ldots, M$. Equation (21.6) now becomes

$$(21.9)\quad -\left.\frac{\partial \bar{f}}{\partial \bar{x}^s}\right|_{\bar{y}} = -\sum_{\alpha=0}^{M} \left.\frac{\partial f}{\partial x^{(\alpha)a}}\right|_{y} X^{\alpha a}_{0s} + \sum_{\rho=0}^{M} \bar{y}_{\rho r} \sum_{\alpha=0}^{M} X^{\rho r}_{\alpha a}\left[X^{\alpha a}_{0s}\right]' .$$

By the excovariance of y we have $y_{\alpha a} = \sum_{\rho=0}^{M} \bar{y}_{\rho r} X^{\rho r}_{\alpha a}$. Since $\left[X^{\alpha a}_{0s}\right]'$ $= X^{(\alpha+1)a}_{0s}$ we are permitted to write (21.9) in the form

$$(21.10) \quad -\left.\frac{\partial \bar{f}}{\partial \bar{x}^s}\right|_{\bar{y}} = -\sum_{\alpha=0}^{M} \left.\frac{\partial f}{\partial x^{(\alpha)a}}\right|_y X^{\alpha a}_{0s} + \sum_{\alpha=0}^{M} y_{\alpha a} X^{(\alpha+1)a}_{0s}$$

$$= -\sum_{\alpha=0}^{M} \left.\frac{\partial f}{\partial x^{(\alpha)a}}\right|_y X^{\alpha a}_{0s} + \sum_{\alpha=1}^{M+1} y_{(\alpha-1)a} X^{\alpha a}_{0s}$$

$$= -\left.\frac{\partial f}{\partial x^a}\right|_y X^{0a}_{0s} + y_{Ma} X^{(M+1)a}_{0s}$$

$$+ \sum_{\alpha=1}^{M} \left[-\left.\frac{\partial f}{\partial x^{(\alpha)a}}\right|_y + y_{(\alpha-1)a} \right] X^{\alpha a}_{0s} .$$

The quantities y_{Ma} transform according to

$$(21.11) \quad \bar{y}_{Ms} = y_{Ma} X^a_s = y_{Ma} X^{(M+1)a}_{(M+1)s} .$$

Let us now examine the transformation formula for $-\left.\frac{\partial f}{\partial x^{(\alpha)a}}\right|_y + y_{(\alpha-1)a}$ for $1 \leq \alpha \leq M$. We have, for $1 \leq \sigma \leq M$, the result

$$(21.12) \quad -\left.\frac{\partial \bar{f}}{\partial \bar{x}^{(\sigma)s}}\right|_{\bar{y}} = -\sum_{\alpha=0}^{M} \frac{\partial f}{\partial x^{(\alpha)a}} \frac{\partial x^{(\alpha)a}}{\partial \bar{x}^{(\sigma)s}} - \sum_{\alpha=0}^{M} \frac{\partial f}{\partial y_{\alpha a}} \frac{\partial y_{\alpha a}}{\partial \bar{x}^{(\sigma)s}} .$$

Toward simplifying (21.12), we use $\frac{\partial y_{\alpha a}}{\partial \bar{x}^{(\sigma)s}} = \frac{\partial}{\partial \bar{x}^{(\sigma)s}} \sum_{\rho=0}^{M} \bar{y}_{\rho r} X^{\rho r}_{\alpha a}$ $= \sum_{\rho=0}^{M} \bar{y}_{\rho r} \frac{\partial}{\partial \bar{x}^{(\sigma)s}} X^{\rho r}_{\alpha a} = \sum_{\rho=0}^{M} \bar{y}_{\rho r} \sum_{\beta=0}^{M} X^{\rho r}_{\alpha a \cdot \beta b} X^{\beta b}_{\sigma s}$ to yield

$$(21.13)\quad \sum_{\alpha=0}^{M} \frac{\partial f}{\partial y_{\alpha a}} \frac{\partial y_{\alpha a}}{\partial \overline{x}^{(\sigma)s}} = \sum_{\alpha=0}^{M} \frac{\partial f}{\partial y_{\alpha a}} \sum_{\rho=0}^{M} \sum_{\beta=0}^{M} \overline{y}_{\rho r} X^{\rho r}_{\alpha a \cdot \beta b} X^{\beta b}_{\sigma s} \quad .$$

In turn, by $\sum_{\beta=0}^{M} X^{\rho r}_{\alpha a \cdot \beta b} X^{\beta b}_{\sigma s} = - \sum_{\beta=0}^{M} \sum_{\tau=0}^{M} X^{\rho r}_{\beta b} X^{\beta b}_{\sigma s \cdot \tau t} X^{\tau t}_{\alpha a}$, we have that (21.13) becomes

$$(21.14)\quad \sum_{\alpha=0}^{M} \frac{\partial f}{\partial y_{\alpha a}} \frac{\partial y_{\alpha a}}{\partial \overline{x}^{(\sigma)s}} = - \sum_{\alpha=0}^{M} \sum_{\beta=0}^{M} \sum_{\rho=0}^{M} \sum_{\tau=0}^{M} \frac{\partial f}{\partial y_{\alpha a}} X^{\tau t}_{\alpha a} \overline{y}_{\rho r} X^{\rho r}_{\beta b} X^{\beta b}_{\sigma s \cdot \tau t} \quad .$$

Also, by $\sum_{\alpha=0}^{M} \frac{\partial f}{\partial y_{\alpha a}} X^{\tau t}_{\alpha a} = \frac{\partial \overline{f}}{\partial \overline{y}_{\tau t}}$ and by $\sum_{\rho=0}^{M} \overline{y}_{\rho r} X^{\rho r}_{\beta b} = y_{\beta b}$ we have that (21.14) becomes

$$(21.15)\quad \sum_{\alpha=0}^{M} \frac{\partial f}{\partial y_{\alpha a}} \frac{\partial y_{\alpha a}}{\partial \overline{x}^{(\sigma)s}} = - \sum_{\beta=0}^{M} \sum_{\tau=0}^{M} y_{\beta b} X^{\beta b}_{\sigma s \cdot \tau t} \frac{\partial \overline{f}}{\partial \overline{y}_{\tau t}} \quad .$$

Now, if (21.8) holds true, we have $\frac{\partial \overline{f}}{\partial \overline{y}_{\tau t}} = \overline{x}^{(\tau+1)t}$ for $\tau = 0, 1, \ldots, M$, so that (21.15) can be replaced by

$$(21.16)\quad - \sum_{\alpha=0}^{M} \frac{\partial f}{\partial y_{\alpha a}} \frac{\partial y_{\alpha a}}{\partial \overline{x}^{(\sigma)s}} = \sum_{\tau=0}^{M} \sum_{\beta=0}^{M} \overline{x}^{(\tau+1)t} y_{\beta b} X^{\beta b}_{\sigma s \cdot \tau t} \quad .$$

By $\sum_{\tau=0}^{M} \overline{x}^{(\tau+1)t} X^{\beta b}_{\sigma s \cdot \tau t} = X^{\beta b\,\prime}_{\sigma s}$, and by $X^{\beta b\,\prime}_{\sigma s} = \binom{\beta}{\sigma} X^{b(\beta-\sigma+1)}_{s} = \binom{\beta}{\sigma}\binom{\beta+1}{\sigma}^{-1} X^{(\beta+1)b}_{\sigma s}$, we have that (21.16) becomes

$$(21.17)\quad - \sum_{\alpha=0}^{M} \frac{\partial f}{\partial y_{\alpha a}} \frac{\partial y_{\alpha a}}{\partial \overline{x}^{(\sigma)s}} = \sum_{\beta=0}^{M} y_{\beta b} \binom{\beta}{\sigma}\binom{\beta+1}{\sigma}^{-1} X^{(\beta+1)b}_{\sigma s} \quad ,$$

or, by the change of the index β of summation to α in the right member of (21.17) according to $\beta = \alpha - 1$, the equivalent

$$(21.18)\quad -\sum_{\alpha=0}^{M} \frac{\partial f}{\partial y_{\alpha a}} \frac{\partial y_{\alpha a}}{\partial \bar{x}_{(\sigma)s}} = \sum_{\alpha=1}^{M+1} y_{(\alpha-1)a}\binom{\alpha-1}{\sigma}\binom{\alpha}{\sigma}^{-1} X^{\alpha a}_{\sigma s} .$$

Then substitution of (21.18) into (21.12) yields a simplified expression for $-\frac{\partial \bar{f}}{\partial \bar{x}_{(\sigma)s}}$ according to

$$(21.19)\quad -\frac{\partial \bar{f}}{\partial \bar{x}_{(\sigma)s}}\bigg|_{\bar{y}} = -\sum_{\alpha=0}^{M} \frac{\partial f}{\partial x_{(\alpha)a}} X^{\alpha a}_{\sigma s} + \sum_{\alpha=1}^{M+1} y_{(\alpha-1)a}\binom{\alpha-1}{\sigma}\binom{\alpha}{\sigma}^{-1} X^{\alpha a}_{\sigma s} .$$

Now let us write a convenient transformation equation for $\bar{y}_{(\sigma-1)s}$ for $1 \leq \sigma \leq M$. We have

$$(21.20)\quad \bar{y}_{(\sigma-1)s} = \sum_{\alpha=0}^{M} y_{\alpha a} X^{\alpha a}_{(\sigma-1)s} = \sum_{\alpha=1}^{M+1} y_{(\alpha-1)a} X^{(\alpha-1)a}_{(\sigma-1)s} = \sum_{\alpha=1}^{M+1} y_{(\alpha-1)a}\binom{\alpha-1}{\sigma-1}\binom{\alpha}{\sigma}^{-1} X^{\alpha a}_{\sigma s}.$$

By (21.19) and (21.20) we have for $1 \leq \sigma \leq M$ the result

$$(21.21)\quad -\frac{\partial \bar{f}}{\partial \bar{x}_{(\sigma)s}}\bigg|_{\bar{y}} + \bar{y}_{(\sigma-1)s} = -\sum_{\alpha=0}^{M} \frac{\partial f}{\partial x_{(\alpha)a}} X^{\alpha a}_{\sigma s} + \sum_{\alpha=1}^{M+1} y_{(\alpha-1)a}\binom{\alpha}{\sigma}^{-1}\left[\binom{\alpha-1}{\sigma} + \binom{\alpha-1}{\sigma-1}\right] X^{\alpha a}_{\sigma s}.$$

Now by $\binom{\alpha}{\sigma}^{-1}\left[\binom{\alpha-1}{\sigma} + \binom{\alpha-1}{\sigma-1}\right] = \binom{\alpha}{\sigma}^{-1}\binom{\alpha}{\sigma} = 1$, we have that (21.21) becomes

$$(21.22)\quad -\frac{\partial \bar{f}}{\partial \bar{x}_{(\sigma)s}}\bigg|_{\bar{y}} + \bar{y}_{(\sigma-1)s} = -\sum_{\alpha=0}^{M} \frac{\partial f}{\partial x_{(\alpha)a}} X^{\alpha a}_{\sigma s} + \sum_{\alpha=1}^{M+1} y_{(\alpha-1)a} X^{\alpha a}_{\sigma s} .$$

By $\frac{\partial f}{\partial x^{(M+1)a}} \equiv 0$, the upper limit M on the first summation in the right member of (21.22) may be replaced by M+1. The term for $\alpha = 0$ can be displayed separately, yielding

$$(21.23) \quad -\sum_{\alpha=0}^{M} \frac{\partial f}{\partial x^{(\alpha)a}} X^{\alpha a}_{\sigma s} = -\frac{\partial f}{\partial x^{a}} X^{0a}_{\sigma s} - \sum_{\alpha=1}^{M+1} \frac{\partial f}{\partial x^{(\alpha)a}} X^{\alpha a}_{\sigma s} .$$

Accordingly, by (21.23), we have that (21.22) becomes, for $1 \leq \sigma \leq M$, the result

$$(21.24) \quad -\left.\frac{\partial \bar{f}}{\partial \bar{x}^{(\sigma)s}}\right|_{\bar{y}} + \bar{y}_{(\sigma-1)s} = -\frac{\partial f}{\partial x^{a}} X^{0a}_{\sigma s} + \sum_{\alpha=1}^{M+1}\left[-\frac{\partial f}{\partial x^{(\alpha)a}} + y_{(\alpha-1)a}\right] X^{\alpha a}_{\sigma s} .$$

By equations (21.10), (21.11), and (21.24) we see that the quantities $F_{\alpha a}$ are the components of an excovariant extensor of range 0 to M+1, where $F_{\alpha a}$ is defined by $F_{0a} = -\left.\frac{\partial f}{\partial x^{a}}\right|_{y}$, $F_{\alpha a} = -\left.\frac{\partial f}{\partial x^{(\alpha)a}}\right|_{y} + y_{(\alpha-1)a}$ for $1 \leq \alpha \leq M$, $F_{(M+1)a} = y_{Ma}$. Thus the assumption that $f(x, x', \ldots, x^{(M)}, y_{\theta a})$ is defined in all coordinate systems by invariance, together with the assumption that $f^{;\beta b} = x'^{(\beta)b}$ along the arc C in question, provides us with a first excovariant extensor associated with $f(x, x', \ldots, x^{(M)}, y_{\theta a})$ and C. A second excovariant extensor $S_{\alpha a}$, of range 0 to M+1, may be constructed from the tensor member of $F_{\alpha a}$ according to Kawaguchi's theorem to yield

$$(21.25) \qquad S_{\alpha a} = \binom{M+1}{\alpha} y_{Ma}{}^{(M+1-\alpha)}$$

for α : 0 to M+1. Accordingly, we now adopt

Definition (21.1). The quantities $F_{\alpha a}$ and $S_{\alpha a}$ with α : 0 to M+1 are called the primary extensors associated with the invariant function $f(x,x', \ldots, x^{(M)}, y_{\theta a})$, y excovariant, and a curve C along which $f^{;\alpha a} = x'^{(\alpha)a}$ for α : 0 to M. Specifically, $F_{0a} = - \left.\frac{\partial f}{\partial x^a}\right|_y$,

$F_{\alpha a} = - \left.\frac{\partial f}{\partial x^{(\alpha)a}}\right|_y + y_{(\alpha-1)a}$ for $1 \leq \alpha \leq M$, $F_{(M+1)a} = y_{Ma}$;

$S_{\alpha a} = \binom{M+1}{\alpha} y_{Ma}{}^{(M+1-\alpha)}$ for $0 \leq \alpha \leq M+1$.

22. <u>A generalized Legendre transform of $f(x,x', \ldots, x^{(M)}, y_{\theta a})$ and its associated primary extensors</u>. We now proceed to construct the primary extensors associated with the function $F(x,x', \ldots, x^{(M)}, x^{(M+1)})$ which is the Legendre transform of $f(x,x', \ldots, x^{(M)}, y_{\theta a})$ with $y_{\theta a} = y_{\theta a}(x,x', \ldots, x^{(M)}, x^{(M+1)})$. The Legendre transform function $F(x,x', \ldots, x^{(M)}, x^{(M+1)})$ is defined by

$$(22.1)\quad F(x,x',\ldots,x^{(M)},x^{(M+1)}) \equiv \sum_{\beta=0}^{M} x'^{(\beta)b} y_{\beta b} - f(x,x',\ldots,x^{(M)},y_{\theta a}).$$

The primary extensors $F_{;\alpha a}$ and $R_{\alpha a}$ (with α : 0 to M+1) are given as usual by $F_{;\alpha a} \equiv \dfrac{\partial F}{\partial x^{(\alpha)a}}$ and by $R_{\alpha a} \equiv \binom{M+1}{\alpha} F_{;(M+1)a}{}^{(M+1-\alpha)}$.

By equation (22.1) we have $F_{;0a} = \dfrac{\partial}{\partial x^{a}} F(x,x',\ldots,x^{(M)},x^{(M+1)})\Big|_{x',\ldots,x^{(M)},x^{(M+1)}}$ $= \displaystyle\sum_{\beta=0}^{M}\sum_{b=1}^{N} x'^{(\beta)b} \frac{\partial y_{\beta b}}{\partial x^{a}} - \frac{\partial f}{\partial x^{a}}\Big|_{y} - \sum_{\beta=0}^{M}\sum_{b=1}^{N} f^{;\beta b} \frac{\partial y_{\beta b}}{\partial x^{a}}$. Because we assume that $f^{;\beta b} = x'^{(\beta)b}$ along the arc C in question, this simplifies to

$$F_{;0a} = -\frac{\partial f}{\partial x^{a}}\Big|_{y}.$$

Also by equation (22.1), for $1 \leq \alpha \leq M$, we have

$$F_{;\alpha a} = \sum_{\beta=0}^{M} y_{\beta b} \frac{\partial x'^{(\beta)b}}{\partial x^{(\alpha)a}} + \sum_{\beta=0}^{M} x'^{(\beta)b} \frac{\partial y_{\beta b}}{\partial x^{(\alpha)a}} - \frac{\partial f}{\partial x^{(\alpha)a}}\Big|_{y}$$

$- \displaystyle\sum_{\beta=0}^{M}\sum_{b=1}^{N} f^{;\beta b} \frac{\partial y_{\beta b}}{\partial x^{(\alpha)a}}$. The second and fourth terms in the right member of this equality combine to vanish. The first term may be simplified according to $\dfrac{\partial x'^{(\beta)b}}{\partial x^{(\alpha)a}} = \delta^{\beta+1}_{\alpha}\,\delta^{b}_{a}$. We have $\displaystyle\sum_{\beta=0}^{M} y_{\beta b} \frac{\partial x'^{(\beta)b}}{\partial x^{(\alpha)a}}$ $= \displaystyle\sum_{\beta=0}^{M} y_{\beta b}\,\delta^{\beta+1}_{\alpha}\,\delta^{b}_{a} = y_{(\alpha-1)a}$ for $1 \leq \alpha \leq M$. Accordingly, for $1 \leq \alpha \leq M$, we have $F_{;\alpha a} = -\dfrac{\partial f}{\partial x^{(\alpha)a}}\Big|_{y} + y_{(\alpha-1)a}$.

Once more by equation (22.1) we have $F_{;(M+1)a} = \displaystyle\sum_{\beta=0}^{M} y_{\beta b} \frac{\partial x'^{(\beta)b}}{\partial x^{(M+1)a}}$ $+ \displaystyle\sum_{\beta=0}^{M} x'^{(\beta)b} \frac{\partial y_{\beta b}}{\partial x^{(M+1)a}} - \sum_{\beta=0}^{M}\sum_{b=1}^{N} f^{;\beta b} \frac{\partial y_{\beta b}}{\partial x^{(M+1)a}}$. Again the assumption that $f^{;\beta b} = x'^{(\beta)b}$ along the arc C in question leads to a simplification, this time to $F_{;(M+1)a} = \sum_{\beta=0}^{M} y_{\beta b} \dfrac{\partial x'^{(\beta)b}}{\partial x^{(M+1)a}}$. But $\dfrac{\partial x'^{(\beta)b}}{\partial x^{(M+1)a}}$ $= \delta^{\beta}_{M}\,\delta^{b}_{a}$. Therefore, $F_{;(M+1)a} = \sum_{\beta=0}^{M} y_{\beta b}\,\delta^{\beta}_{M}\,\delta^{b}_{a} = y_{Ma}$.

We have now shown by differentiation of equation (22.1) that $F_{;\alpha a}$ is given for α : 0 to M+1 by $F_{;0a} = -\left.\dfrac{\partial f}{\partial x^{a}}\right|_{y}$; $F_{;\alpha a} = -\left.\dfrac{\partial f}{\partial x^{(\alpha)a}}\right|_{y} + y_{(\alpha-1)a}$ for $1 \leq \alpha \leq M$; and $F_{;(M+1)a} = y_{Ma}$. Furthermore, because $R_{\alpha a}$ is defined for α : 0 to M+1 to have the Kawaguchi structure based on the tensor rank $\alpha = M+1$ of $F_{;\alpha a}$, we have $R_{\alpha a} = \binom{M+1}{\alpha} y_{Ma}^{\;(M+1-\alpha)}$ for α : 0 to M+1. Thus we have

Theorem (22.1). The primary extensors $F_{;\alpha a}$ and $R_{\alpha a}$ (with α : 0 to M+1) associated with the Legendre transform function $F(x,x',\ldots,x^{(M)},x^{(M+1)})$ of $f(x,x',\ldots,x^{(M)},y_{\theta a})$, given by

$$F(x,x',\ldots,x^{(M)},x^{(M+1)}) = \sum_{\beta=0}^{M} x'^{(\beta)b} y_{\beta b} - f(x,x',\ldots,x^{(M)},y_{\theta a}),$$

are equal, respectively, to the primary extensors $F_{\alpha a}$ and $S_{\alpha a}$ associated with $f(x,x',\ldots,x^{(M)},y_{\theta a})$.

23. <u>A generalization of the concept of primary extensor associated with a function $f(x, x', \ldots, x^{(M)}, y_{\theta a})$.</u> In Section 12 (see equation (12.8)) there was given a generalization of the concept of primary extensor such as to apply to a class C^{M+1} function $F(x, x', \ldots, x^{(M)})$ defined along a class C^{2M} parameterized arc. This involved a result obtained by constructing from $F(x, x', \ldots, x^{(M)})$ not only (1) a gradient extensor $0_{\alpha a}(F_{;\alpha a})$, but also (2) a set of M additional extensors $L_{\alpha a}$ (with L = 1, 2, . . . , M) such that $L_{0a} = F_{;La}{}^{(L)}$ for each L such that $0 \leq L \leq M$. It is our purpose in this section to construct a somewhat analogous generalization of the concept of primary extensor associated with a class C^{M+1} function $f(x, x', \ldots, x^{(M)}, y_{\theta a})$ defined over a class C^{2M} parameterized arc along which is valid the extensor equation $\left.\frac{\partial f}{\partial y_{\beta b}}\right|_{x} = x'^{(\beta)b}$. We note in passing the probable existence of generalizations other than the forthcoming.

Assuming that the first primary extensor $F_{\alpha a}$ of Definition (21.1) plays the role of a zeroth member in a set of generalized primary extensors, as did $F_{;\alpha a}$ when $F(x,x', \ldots, x^{(M)})$ was the base function, we seek a set $\{\phi_{\alpha a} | 0 \leq \phi \leq M\}$ of primary extensors associated with $f(x,x', \ldots, x^{(M)}, y_{\theta a})$ such that, for $0 \leq \phi \leq M$, we have

$$(23.1) \qquad \phi_{0a} = - \left. \frac{\partial f}{\partial x^{(\phi)a}} \right|_y .$$

We will use the extensor transformation law to determine from (23.1) how the nonzero ranks of the extensors $\phi_{\alpha a}$ must be defined; we will then prove that, for $0 \leq \phi \leq M$, the quantities $\phi_{\alpha a}$ so defined are extensor components.

Accordingly, we begin by examining a transformation equation for $- \left. \frac{\partial f}{\partial x^{(\phi)a}} \right|_y$. In an x-coordinate system such an equation follows from (21.19) by replacing σ by ϕ to obtain the transformation equation for $- \left. \frac{\partial \bar{f}}{\partial \bar{x}^{(\phi)s}} \right|_{\bar{y}}$ given as

$$(23.2) \qquad - \left. \frac{\partial \bar{f}}{\partial \bar{x}^{(\phi)s}} \right|_{\bar{y}} = - \sum_{\alpha=\phi}^{M} \left. \frac{\partial f}{\partial x^{(\alpha)a}} \right|_y X^{\alpha a}_{\phi s} + \sum_{\alpha=\phi+1}^{M+1} y_{(\alpha-1)a} \binom{\alpha-1}{\phi} \binom{\alpha}{\phi}^{-1} X^{\alpha a}_{\phi s} .$$

Because $X^{\alpha a}_{\phi s} \equiv 0$ for $\alpha < \phi$, the lower limit on the first summation in the right member of (23.2) is ϕ, rather than zero. A change of index in the derivation of the second summation in (23.2) caused the change in the summation limits. Now by substitution of $\binom{\alpha}{\phi} X^{a(\alpha-\phi)}_{s}$ for $X^{\alpha a}_{\phi s}$, and by substitution of $X^{a(\alpha-\phi)}_{s}$ for $\binom{\alpha}{\phi}^{-1} X^{\alpha a}_{\phi s}$ in (23.2), we have

$$(23.3)\quad -\left.\frac{\partial \bar{f}}{\partial \bar{x}^{(\phi)s}}\right|_{\bar{y}} = -\sum_{\alpha=\phi}^{M} \left.\frac{\partial f}{\partial x^{(\alpha)a}}\right|_{y} \binom{\alpha}{\phi} X^{a(\alpha-\phi)}_{s} + \sum_{\alpha=\phi+1}^{M+1} y_{(\alpha-1)a} \binom{\alpha-1}{\phi} X^{a(\alpha-\phi)}_{s} .$$

Replacing the index α of summation by $\alpha + \phi$ in both summations in the right member of (23.3), we obtain

$$(23.4)\quad -\left.\frac{\partial \bar{f}}{\partial \bar{x}^{(\phi)s}}\right|_{\bar{y}} = -\sum_{\alpha=0}^{M-\phi} \left.\frac{\partial f}{\partial x^{(\alpha+\phi)a}}\right|_{y} \binom{\alpha+\phi}{\phi} X^{a(\alpha)}_{s} + \sum_{\alpha=1}^{M+1-\phi} y_{(\alpha+\phi-1)a} \binom{\alpha+\phi-1}{\phi} X^{a(\alpha)}_{s} .$$

Because $\left.\frac{\partial f}{\partial x^{(M+1)a}}\right|_{y} \equiv 0$, the upper limit on the first summation in the right member of (23.4) may be elevated to $M + 1 - \phi$. Also, because

$\binom{\alpha+\phi-1}{\phi} \equiv 0$ if $\alpha < 1$, the lower limit on the second summation in the right member of (23.4) may be lowered to zero, provided $y_{(\phi-1)a}$ is defined to have some arbitrary value, say zero, in case $\phi = 0$. With this provision and the indicated alterations in ranges of summation, the two terms in the right member of (23.4) may be combined into one to obtain

$$(23.5)\quad -\left.\frac{\partial\bar{f}}{\partial\bar{x}^{(\phi)s}}\right|_{\bar{y}} = \sum_{\alpha=0}^{M+1-\phi}\left[-\left.\frac{\partial f}{\partial x^{(\alpha+\phi)a}}\right|_{y}\binom{\alpha+\phi}{\phi} + y_{(\alpha+\phi-1)a}\binom{\alpha+\phi-1}{\phi}\right]X_s^{a(\alpha)}.$$

We now replace the left member of (23.5) by the symbol $\bar{\phi}_{0s}$ to emphasize that we have a transformation equation for what is to be the zero rank of an extensor. We also replace $X_s^{a(\alpha)}$ in the right member of (23.5) by $X_{0s}^{\alpha a}$ to obtain, for $0 \leq \phi \leq M$, the result

$$(23.6)\quad \bar{\phi}_{0s} = \sum_{\alpha=0}^{M+1-\phi}\left[-\left.\frac{\partial f}{\partial x^{(\alpha+\phi)a}}\right|_{y}\binom{\alpha+\phi}{\phi} + y_{(\alpha+\phi-1)a}\binom{\alpha+\phi-1}{\phi}\right]X_{0s}^{\alpha a}\ .$$

Equation (23.6) suggests, but does not prove, Theorem (23.1) regarding the generalized primary extensors $\phi_{\alpha a}$ for which $0 \leq \phi \leq M$.

Theorem (23.1). Given (1) the class c^{M+1} invariant function $f(x, x', \ldots x^{(M)}, y_{\theta a})$ of weight zero with y excovariant, and (2) the class C^{2M} arc C, along which the extensor equation $f^{;\beta b} = x'^{(\beta)b}$ is valid, where $f^{;\beta b} \overset{\text{def}}{\equiv} \left.\frac{\partial f}{\partial y_{\beta b}}\right|_x$. If ϕ satisfies $0 \leq \phi \leq M$, then the quantities $\phi_{\alpha a}$ given by

$$\phi_{\alpha a} = -\binom{\alpha+\phi}{\phi} \left.\frac{\partial f}{\partial x^{(\alpha+\phi)a}}\right|_y + \binom{\alpha+\phi-1}{\phi} y_{(\alpha+\phi-1)a}$$

are the components of an extensor of range α : 0 to $M + 1 - \phi$.

Proof: We wish to show that, for any fixed ϕ such that $0 \leq \phi \leq M$, if

$$\overline{\phi}_{\sigma s} = -\binom{\sigma+\phi}{\phi} \left.\frac{\partial \overline{f}}{\partial \overline{x}^{(\sigma+\phi)s}}\right|_{\overline{y}} + \binom{\sigma+\phi-1}{\phi} \overline{y}_{(\sigma+\phi-1)s}$$

for each σ such that $0 \leq \sigma \leq M + 1 - \phi$, then

$$\overline{\phi}_{\sigma s} = \sum_{\alpha=0}^{M+1-\phi} \phi_{\alpha a} X^{\alpha a}_{\sigma s}$$

for each σ such that $0 \leq \sigma \leq M + 1 - \phi$.

In fact we have, for $0 \leq \sigma + \phi \leq M$ by the chain rule of differentiation, the result

$$(23.7) \quad \frac{\partial \overline{f}}{\partial \overline{x}^{(\sigma+\phi)s}} = \sum_{\alpha=0}^{M} \left.\frac{\partial f}{\partial x^{(\alpha)a}}\right|_y \frac{\partial x^{(\alpha)a}}{\partial \overline{x}^{(\sigma+\phi)s}} + \sum_{\alpha=0}^{M} \frac{\partial f}{\partial y_{\alpha a}} \frac{\partial y_{\alpha a}}{\partial \overline{x}^{(\sigma+\phi)s}} .$$

Now, because $y_{\alpha a} = \sum_{\rho=0}^{M} \bar{y}_{\rho r} X^{\rho r}_{\alpha a}$, we have

$$(23.8) \qquad \frac{\partial y_{\alpha a}}{\partial \bar{x}^{(\sigma+\phi)s}} = \sum_{\rho=0}^{M} \sum_{\beta=0}^{M} \bar{y}_{\rho r} X^{\rho r}_{\alpha a \cdot \beta b} X^{\beta b}_{(\sigma+\phi)s} .$$

Because $0 = \left[\delta^{\rho}_{\sigma+\phi} \delta^{r}_{s} \right]_{;\alpha a} = \left[\sum_{\beta=0}^{M} X^{\rho r}_{\beta b} X^{\beta b}_{(\sigma+\phi)s} \right]_{;\alpha a} = \sum_{\beta=0}^{M} X^{\rho r}_{\alpha a \cdot \beta b} X^{\beta b}_{(\sigma+\phi)s}$
$+ \sum_{\beta=0}^{M} \sum_{\tau=0}^{M} X^{\rho r}_{\beta b} X^{\beta b}_{(\sigma+\phi)s \cdot \tau t} X^{\tau t}_{\alpha a}$, we have

$$(23.9) \qquad \sum_{\beta=0}^{M} X^{\rho r}_{\alpha a \cdot \beta b} X^{\beta b}_{(\sigma+\phi)s} = - \sum_{\beta=0}^{M} \sum_{\tau=0}^{M} X^{\rho r}_{\beta b} X^{\beta b}_{(\sigma+\phi)s \cdot \tau t} X^{\tau t}_{\alpha a} .$$

Accordingly, (23.8) becomes

$$(23.10) \qquad \frac{\partial y_{\alpha a}}{\partial \bar{x}^{(\sigma+\phi)s}} = - \sum_{\rho=0}^{M} \sum_{\beta=0}^{M} \sum_{\tau=0}^{M} \bar{y}_{\rho r} X^{\rho r}_{\beta b} X^{\beta b}_{(\sigma+\phi)s \cdot \tau t} X^{\tau t}_{\alpha a},$$

from which there follows

$$(23.11) \qquad \sum_{\alpha=0}^{M} \frac{\partial f}{\partial y_{\alpha a}} \frac{\partial y_{\alpha a}}{\partial \bar{x}^{(\sigma+\phi)s}} = - \sum_{\alpha=0}^{M} \sum_{\rho=0}^{M} \sum_{\beta=0}^{M} \sum_{\tau=0}^{M} \frac{\partial f}{\partial y_{\alpha a}} \bar{y}_{\rho r} X^{\rho r}_{\beta b} X^{\beta b}_{(\sigma+\phi)s \cdot \tau t} X^{\tau t}_{\alpha a} .$$

Now $\sum_{\alpha=0}^{M} \frac{\partial f}{\partial y_{\alpha a}} X^{\tau t}_{\alpha a} = \frac{\partial \bar{f}}{\partial \bar{y}_{\tau t}}$, and $\sum_{\rho=0}^{M} \bar{y}_{\rho r} X^{\rho r}_{\beta b} = y_{\beta b}$. Therefore, (23.11) simplifies to

$$(23.12) \qquad \sum_{\alpha=0}^{M} \frac{\partial f}{\partial y_{\alpha a}} \frac{\partial y_{\alpha a}}{\partial \bar{x}^{(\sigma+\phi)s}} = - \sum_{\beta=0}^{M} \sum_{\tau=0}^{M} \frac{\partial \bar{f}}{\partial \bar{y}_{\tau t}} y_{\beta b} X^{\beta b}_{(\sigma+\phi)s \cdot \tau t} .$$

Next, we impose the requirement that, along the arc C in question, the extensor equation $\left.\frac{\partial f}{\partial y_{\alpha a}}\right|_{x} = x^{'(\alpha)a}$ holds true. Accordingly, $\frac{\partial \bar{f}}{\partial \bar{y}_{\tau t}} = \bar{x}^{(\tau+1)t}$, and (23.12) becomes

$$(23.13)\quad \sum_{\alpha=0}^{M} \frac{\partial f}{\partial y_{\alpha a}} \frac{\partial y_{\alpha a}}{\partial \bar{x}^{(\sigma+\phi)s}} = -\sum_{\beta=0}^{M} \sum_{\tau=0}^{M} \bar{x}^{(\tau+1)t} y_{\beta b} X^{\beta b}_{(\sigma+\phi)s\cdot\tau t} .$$

But $\sum_{\tau=0}^{M} \bar{x}^{(\tau+1)t} X^{\beta b}_{(\sigma+\phi)s\cdot\tau t} = X^{\beta b}_{(\sigma+\phi)s}{}'$, and $X^{\beta b}_{(\sigma+\phi)s}{}' = \binom{\beta}{\sigma+\phi} X_s^{b(\beta-\sigma-\phi+1)}$.

Hence, (23.13) can be written in the form

$$(23.14)\quad \sum_{\alpha=0}^{M} \frac{\partial f}{\partial y_{\alpha a}} \frac{\partial y_{\alpha a}}{\partial \bar{x}^{(\sigma+\phi)s}} = -\sum_{\beta=0}^{M} y_{\beta b} \binom{\beta}{\sigma+\phi} X_s^{b(\beta-\sigma-\phi+1)} .$$

The index of summation β in the right member of (23.14) may be replaced by α according to the rule $\beta + 1 = \alpha$. The right member of (23.14) becomes $-\sum_{\alpha=1}^{M+1} y_{(\alpha-1)a} \binom{\alpha-1}{\sigma+\phi} X_s^{b(\alpha-\sigma-\phi)}$. By $X_s^{b(\alpha-\sigma-\phi)} = \binom{\alpha}{\sigma+\phi}^{-1} X^{\alpha a}_{(\sigma+\phi)s}$ there follows another alteration of the right member of (23.14), yielding

$$(23.15)\quad \sum_{\alpha=0}^{M} \frac{\partial f}{\partial y_{\alpha a}} \frac{\partial y_{\alpha a}}{\partial \bar{x}^{(\sigma+\phi)s}} = -\sum_{\alpha=1}^{M+1} y_{(\alpha-1)a} \binom{\alpha-1}{\sigma+\phi} \binom{\alpha}{\sigma+\phi}^{-1} X^{\alpha a}_{(\sigma+\phi)s} .$$

Equation (23.15) may be substituted into (23.7) to give

$$(23.16)\quad \frac{\partial \bar{f}}{\partial \bar{x}^{(\sigma+\phi)s}} = \sum_{\alpha=0}^{M} \left.\frac{\partial f}{\partial x^{(\alpha)a}}\right|_{y} X^{\alpha a}_{(\sigma+\phi)s} - \sum_{\alpha=1}^{M+1} y_{(\alpha-1)a} \binom{\alpha-1}{\sigma+\phi} \binom{\alpha}{\sigma+\phi}^{-1} X^{\alpha a}_{(\sigma+\phi)s} .$$

We may utilize the identity $\frac{\partial f}{\partial x^{(M+1)a}} \equiv 0$ to elevate the upper limit on the first summation in the right member of (23.16) to M+1. We may also invoke a definition assigning the value zero to the symbols $y_{(\alpha-1)a}\binom{\alpha-1}{\sigma+\phi}\binom{\alpha}{\sigma+\phi}^{-1}$ for $\alpha = 0$, to lower the lower limit on the second summation in the right member of (23.16) to 0. Then, by multiplication of both members of (23.16) by $\binom{\sigma+\phi}{\phi}$, we have

$$(23.17)\quad \binom{\sigma+\phi}{\phi}\frac{\partial \bar{f}}{\partial \bar{x}^{(\sigma+\phi)s}}\bigg|_{\bar{y}} = \sum_{\alpha=0}^{M+1} \frac{\partial f}{\partial x^{(\alpha)a}}\bigg|_{y}\binom{\sigma+\phi}{\phi} X^{\alpha a}_{(\sigma+\phi)s}$$

$$- \sum_{\alpha=0}^{M+1} y_{(\alpha-1)a}\binom{\alpha-1}{\sigma+\phi}\binom{\sigma+\phi}{\phi}\binom{\alpha}{\sigma+\phi}^{-1} X^{\alpha a}_{(\sigma+\phi)s}$$

for σ : 0 to M - ϕ. The range of values of σ for which (23.17) is true may be extended to σ : 0 to M - ϕ + 1, as the case σ = M - ϕ + 1 produces the identity 0 = 0.

Similarly, by multiplication of both members of an extensor transformation formula for $\bar{y}_{(\sigma+\phi-1)s}$ by $\binom{\sigma+\phi-1}{\phi}$, we have

$$(23.18)\quad \binom{\sigma+\phi-1}{\phi}\bar{y}_{(\sigma+\phi-1)s} = \binom{\sigma+\phi-1}{\phi}\sum_{\alpha=0}^{M} y_{\alpha a} X^{\alpha a}_{(\sigma+\phi-1)s}$$

for $\sigma : 0$ to $M - \phi + 1$, where the definitions $\binom{-1}{0} \overset{\text{def}}{\equiv} 0$, $\bar{y}_{(-1)s} \overset{\text{def}}{\equiv} 0$, $X^{0a}_{(-1)s} \overset{\text{def}}{\equiv} 0$ are invoked for the case $\phi = \sigma = 0$. Now, by $X^{a(\gamma)}_{s} \overset{\text{def}}{\equiv} 0$ for $\gamma < 0$ and by $X^{\alpha a}_{(\sigma+\phi-1)s} = \binom{\alpha}{\sigma+\phi-1} X^{a(\alpha-\sigma-\phi+1)}_{s}$, the right member of (23.18) becomes $\sum_{\alpha=0}^{M} y_{\alpha a} \binom{\sigma+\phi-1}{\phi} \binom{\alpha}{\sigma+\phi-1} X^{a(\alpha-\sigma-\phi+1)}_{s}$. By a replacement of the index α of summation by $\alpha - 1$ this new right member of (23.18) can be written $\sum_{\alpha=1}^{M+1} y_{(\alpha-1)a} \binom{\sigma+\phi-1}{\phi} \binom{\alpha-1}{\sigma+\phi-1} X^{a(\alpha-\sigma-\phi)}_{s}$. By $y_{(\alpha-1)a} \cdot \binom{\sigma+\phi-1}{\phi} \binom{\alpha-1}{\sigma+\phi-1} \overset{\text{def}}{\equiv} 0$ for $\alpha = 0$, and for $\sigma = \phi = 0$ the lower limit on this new α-summation may be lowered to zero. Also, $\binom{\alpha-1}{\sigma+\phi-1} \equiv 0$ for $\alpha < 1$. The symbol $X^{a(\alpha-\sigma-\phi)}_{s}$ may be replaced by $\binom{\alpha}{\sigma+\phi}^{-1} X^{\alpha a}_{(\sigma+\phi)s}$, where $\binom{0}{\sigma+\phi}^{-1} \overset{\text{def}}{\equiv} 0$. We obtain

$$(23.19) \qquad \binom{\sigma+\phi-1}{\phi} \bar{y}_{(\sigma+\phi-1)s} = \sum_{\alpha=0}^{M+1} y_{(\alpha-1)a} \binom{\sigma+\phi-1}{\phi} \binom{\alpha-1}{\sigma+\phi-1} \binom{\alpha}{\sigma+\phi}^{-1} X^{\alpha a}_{(\sigma+\phi)s}.$$

Now the structure for $\bar{\phi}_{\sigma s}$ for $\sigma : 0$ to $M - \phi + 1$ assumed as a hypothesis of our Theorem (23.1) is that of (23.20), namely

$$(23.20) \qquad \bar{\phi}_{\sigma s} = - \binom{\sigma+\phi}{\phi} \frac{\partial \bar{f}}{\partial \bar{x}_{(\sigma+\phi)s}} + \binom{\sigma+\phi-1}{\phi} \bar{y}_{(\sigma+\phi-1)s} \; .$$

By (23.17) and (23.19) we get, with our special symbol definitions for the case $\alpha = 0$, the result

$$(23.21)\quad \bar{\phi}_{\sigma s} = -\sum_{\alpha=0}^{M+1} \frac{\partial f}{\partial x_{(\alpha)a}}\bigg|_y \binom{\sigma+\phi}{\phi} X^{\alpha a}_{(\sigma+\phi)s}$$

$$+\sum_{\alpha=0}^{M+1} y_{(\alpha-1)a}\binom{\alpha-1}{\sigma+\phi}\binom{\sigma+\phi}{\phi}\binom{\alpha}{\sigma+\phi}^{-1} X^{\alpha a}_{(\sigma+\phi)s}$$

$$+\sum_{\alpha=0}^{M+1} y_{(\alpha-1)a}\binom{\sigma+\phi-1}{\phi}\binom{\alpha-1}{\sigma+\phi-1}\binom{\alpha}{\sigma+\phi}^{-1} X^{\alpha a}_{(\sigma+\phi)s} \quad .$$

The expression $\binom{\alpha-1}{\sigma+\phi}\binom{\sigma+\phi}{\phi}$ in the second summation in the right member of (23.21) may be replaced by $\binom{\alpha-1}{\phi}\binom{\alpha-1-\phi}{\sigma}$. The expression $\binom{\sigma+\phi-1}{\phi}\binom{\alpha-1}{\sigma+\phi-1}$ in the third summation in the right member of (23.21) may be replaced by $\binom{\alpha-1}{\phi}\binom{\alpha-1-\phi}{\sigma-1}$. The second and third summations in the right member of (23.21) may now be combined, using $\binom{\alpha-1}{\phi}\left[\binom{\alpha-1-\phi}{\sigma}+\binom{\alpha-1-\phi}{\sigma-1}\right]=\binom{\alpha-1}{\phi}\binom{\alpha-\phi}{\sigma}$. By $y_{(\alpha-1)a}\binom{\alpha}{\sigma+\phi}^{-1}\binom{\alpha-1}{\phi}\binom{\alpha-\phi}{\sigma} \overset{\text{def}}{\equiv} 0$ for $\alpha = 0$, there results

$$(23.22)\quad \bar{\phi}_{\sigma s} = -\sum_{\alpha=0}^{M+1} \frac{\partial f}{\partial x_{(\alpha)a}}\bigg|_y \binom{\sigma+\phi}{\phi} X^{\alpha a}_{(\sigma+\phi)s}$$

$$+\sum_{\alpha=0}^{M+1} y_{(\alpha-1)a}\binom{\alpha}{\sigma+\phi}^{-1}\binom{\alpha-1}{\phi}\binom{\alpha-\phi}{\sigma} X^{\alpha a}_{(\sigma+\phi)s} \quad .$$

We now combine the two summations in the right member of (23.22) into a single summation, where the fact that $X^{\alpha a}_{(\sigma+\phi)s} \equiv 0$ for $\alpha < \sigma + \phi$

allows us to raise the lower limit on the new α-summation from 0 to ϕ. We have

$$(23.23)\quad \overline{\phi}_{\sigma s} = \sum_{\alpha=\phi}^{M+1}\left[-\frac{\partial f}{\partial x^{(\alpha)a}}\bigg|_y \binom{\sigma+\phi}{\phi} + y_{(\alpha-1)a}\binom{\alpha}{\sigma+\phi}^{-1}\binom{\alpha-1}{\phi}\binom{\alpha-\phi}{\sigma}\right] X^{\alpha a}_{(\sigma+\phi)s}\,.$$

Next, we utilize the identity $X^{\alpha a}_{(\sigma+\phi)s} = \binom{\alpha}{\sigma+\phi}\binom{\alpha-\phi}{\sigma}^{-1} X^{(\alpha-\phi)a}_{\sigma s}$, where $\binom{\alpha-\phi}{\sigma}^{-1} X^{(\alpha-\phi)a}_{\sigma s} \overset{\text{def}}{\equiv} 0$ for $\alpha < \phi$, to rewrite (23.13) according to

(23.24)

$$\overline{\phi}_{\sigma s} = \sum_{\alpha=\phi}^{M+1}\left[-\frac{\partial f}{\partial x^{(\alpha)a}}\bigg|_y \binom{\sigma+\phi}{\phi} + y_{(\alpha-1)a}\binom{\alpha}{\sigma+\phi}^{-1}\binom{\alpha-1}{\phi}\binom{\alpha-\phi}{\sigma}\right]\binom{\alpha}{\sigma+\phi}\binom{\alpha-\phi}{\sigma}^{-1} X^{(\alpha-\phi)a}_{\sigma s}\,.$$

Then we replace the summation index α in (23.24) by $\alpha + \phi$ to yield

$$(23.25)\quad \overline{\phi}_{\sigma s} = \sum_{\alpha=0}^{M+1-\phi}\left[-\frac{\partial f}{\partial x^{(\alpha+\phi)a}}\bigg|_y \binom{\sigma+\phi}{\phi} + y_{(\alpha+\phi-1)a}\binom{\alpha+\phi}{\sigma+\phi}^{-1}\binom{\alpha+\phi-1}{\phi}\binom{\alpha}{\sigma}\right]$$

$$\binom{\alpha+\phi}{\sigma+\phi}\binom{\alpha}{\sigma}^{-1} X^{\alpha a}_{\sigma s}\,.$$

Finally, because $\binom{\sigma+\phi}{\phi}\binom{\alpha+\phi}{\sigma+\phi}\binom{\alpha}{\sigma}^{-1} = \binom{\alpha+\phi}{\phi}$, and because

$\binom{\alpha+\phi}{\sigma+\phi}^{-1}\binom{\alpha+\phi-1}{\phi}\binom{\alpha}{\sigma}\binom{\alpha+\phi}{\sigma+\phi}\binom{\alpha}{\sigma}^{-1} = \binom{\alpha+\phi}{\sigma+\phi}^{-1}\binom{\alpha+\phi-1}{\phi}\binom{\alpha+\phi}{\sigma+\phi} = \binom{\alpha+\phi-1}{\phi}$, we have

that (23.25) becomes

$$(23.26)\quad \overline{\phi}_{\sigma s} = -\sum_{\alpha=0}^{M+1-\phi}\frac{\partial f}{\partial x^{(\alpha+\phi)a}}\bigg|_y \binom{\alpha+\phi}{\phi} X^{\alpha a}_{\sigma s} + \sum_{\alpha=0}^{M+1-\phi} y_{(\alpha+\phi+1)a}\binom{\alpha+\phi-1}{\phi} X^{\alpha a}_{\sigma s}$$

for σ : 0 to $M - \phi + 1$.

Accordingly, for σ : 0 to $M - \phi + 1$, we have

$$(23.27)\quad \bar{\phi}_{\sigma s} = \sum_{\alpha=0}^{M+1-\phi} \left[-\binom{\alpha+\phi}{\phi} \left.\frac{\partial f}{\partial x^{(\alpha+\phi)a}}\right|_y + \binom{\alpha+\phi-1}{\phi} y_{(\alpha+\phi-1)a} \right] X^{\alpha a}_{\sigma s}$$

for any fixed ϕ such that $0 \leq \phi \leq M$, i.e., for σ : 0 to $M - \phi + 1$, we have

$$(23.28)\quad \bar{\phi}_{\sigma s} = \sum_{\alpha=0}^{M+1-\phi} \phi_{\alpha a} X^{\alpha a}_{\sigma s}$$

for any fixed ϕ such that $0 \leq \phi \leq M$, and Theorem (23.1) is proved.

Now for each ϕ such that $0 \leq \phi \leq M$ we have defined an extensor $\phi_{\alpha a}$ of range α : 0 to $M + 1 - \phi$. As we have already indicated, it seems natural to regard our set $\{\phi_{\alpha a} | 0 \leq \phi \leq M\}$ as one set of generalized primary extensors associated with the function $f(x, x', \ldots, x^{(M)}, y_{\theta a})$. However, the results of Chapter IV will suggest to us that our set $\{\phi_{\alpha a} | 0 \leq \phi \leq M\}$ is but a proper subset of a larger set $\mathcal{A}(E_{\alpha a})$, $E_{\alpha a} \overset{\text{def}}{\equiv} \phi_{\alpha a}$ for $\phi = 0$, of fundamental extensors associated with $E_{\alpha a}$. Accordingly, a topic for further research would then be to utilize $\mathcal{S}(E_{\alpha a})$ to obtain a treatment of Hamiltonian dynamics which generalizes the methods of Section 24.

We now terminate Section 23 by defining our particular set $\{\phi_{\alpha a} | 0 \leq \phi \leq M\}$ of generalized primary extensors associated with the function $f(x, x', \ldots, x^{(M)}, y_{\theta a})$ and an arc C along which $\left.\frac{\partial f}{\partial y_{\beta b}}\right|_x = x'^{(\beta)b}$:

Definition (23.1). Given (1) the class C^{M+1} invariant function $f(x, x', \ldots, x^{(M)}, y_{\theta a})$ of weight zero with y excovariant, and (2) the class C^{2M} arc C, along which the extensor equation $f^{;\beta b} = x'^{(\beta)b}$ is valid, where $f^{;\beta b} \overset{\text{def}}{\equiv} \left.\frac{\partial f}{\partial y_{\beta b}}\right|_x$. The set $\{\phi_{\alpha a} | 0 \leq \phi \leq M\}$ of generalized primary extensors associated with f and C consists of the M+1 extensors $\phi_{\alpha a}$ of range α : 0 to $M + 1 - \phi$ given by

$$\phi_{\alpha a} = -\binom{\alpha+\phi}{\phi} \left.\frac{\partial f}{\partial x^{(\alpha+\phi)a}}\right|_y + \binom{\alpha+\phi-1}{\phi} y_{(\alpha+\phi-1)a}$$

for ϕ any fixed value such that $0 \leq \phi \leq M$.

24. Primary extensors associated with the differentiated total energy function $[h(q,p)]^{(M)}$ and the Hamiltonian equations of motion. We now concern ourselves with the dynamical system T(q,q'), V(q). Here, the q's are space coordinates, the potential energy function V(q) depends upon the q's alone, and the kinetic energy function T(q,q') is given by a quadratic form in the q' 's, namely $T = \frac{1}{2} g_{ab} q^{'a} q^{'b}$. The independent variable t is time, and the metric tensor g_{ab} is given by $g_{ab} = \rho_a \cdot \rho_b$, where ρ is the radius vector and $\rho_a = \frac{\partial \rho}{\partial q^a}, \rho_b = \frac{\partial \rho}{\partial q^b}$.

If we define generalized momentum coordinates p_a by $p_a \equiv T_{;1a}$ so that $p_a = g_{ab} q^{'b}$, it is possible to express a total energy function h(q,p) in Hamiltonian form, $h(q,p) = t(q,p) + V(q)$. Because the quantities p_a may be regarded as the covariant description of the $q^{'a}$ relative to g_{ab} for our mechanical system according to $p_a = g_{ab} q^{'b}$, $q^{'a} = g^{ab} p_b$, where $g^{ab} \equiv \frac{c_{ab}}{G}$, with c_{ab} the cofactor of g_{ab} and G the determinant of the g_{ab}'s, we have a kinetic energy function t(q,p) replacing T(q,q') via $T(q,q') = \frac{1}{2} g_{ab} q^{'a} q^{'b} = \frac{1}{2} g_{ab} g^{ac} p_c g^{bd} p_d = \frac{1}{2} \delta^c_b p_c g^{bd} p_d = \frac{1}{2} p_b p_d g^{bd} = t(q,p)$. The space coordinates q and the momentum coordinates p are treated as independent variables in phase space and in Hamiltonian mechanics.

Now it was noted in Section 14 (see also [2], p. 62) that the equations of motion of Lagrangian dynamics may be expressed as the equality of the primary extensors associated with $\left[L(q,q')\right]^{(Q)}$, where Q is an arbitrary positive integer and where $L(q,q') = T(q,q') - V(q)$, assuming (1) that the principle of conservation of energy holds, and (2) that the differential equations of motion are of order two or more. Also noted in Section 14 (see also [2]) was the result that the introduction of the tensor equation $\frac{\partial h}{\partial p_a} = q'^a$ permits the expression of the equivalent Hamiltonian equations as the equality of a pair of primary extensors associated with h(q,p). This alternate formulation of the dynamical problem was made obvious by noting in Section 14 (and [2], p. 62) that the functions L(q,q') and h(q,p) are related in the manner of the Legendre transformation in Theorem (14.1), so that the respective primary extensors associated with these functions are identical. Finally, in Theorem (22.1), we saw that the primary extensors associated with the generalized Legendre transform function $F(x,x', \ldots, x^{(M)}, x^{(M+1)})$ of $f(x,x', \ldots, x^{(M)}, y_{\theta a})$ are equal to the respective primary extensors associated with f. These results suggest that the Hamiltonian equations and their derivatives may be expressible as the equality of the primary extensors associated with the differentiated total energy function $\left[h(q,p)\right]^{(M)}$, M an arbitrary positive integer. Before we investigate this possibility, let us first compare two extensors (of different ranges) associated with $T^{(M)}$.

The gradient extensor $G_{\beta a}, G_{(\alpha+1)a} \overset{\text{def}}{\equiv} \dfrac{\partial}{\partial q'^{(\alpha)a}} T^{(M)}$, of range β : 1 to M+1 is similar to the Kawaguchi extensor $K_{\alpha a}$, $K_{\alpha a} \overset{\text{def}}{\equiv} \binom{M}{\alpha} p_a^{(M-\alpha)}$ of range α : 0 to M based on the momentum tensor p_a. This results as follows: From $T = \frac{1}{2} g_{cd} q'^c q'^d$, obtain by the Leibnitz differentiation formula the result

$$(24.1) \qquad T^{(M)} = \frac{1}{2} \sum_{\mu=0}^{M} \binom{M}{\mu} g_{cd}^{(M-\mu)} \left(q'^c q'^d\right)^{(\mu)} .$$

By a further application of the Leibnitz formula there follows

$$(24.2) \qquad T^{(M)} = \frac{1}{2} \sum_{\mu=0}^{M} \binom{M}{\mu} g_{cd}^{(M-\mu)} \sum_{\beta=0}^{\mu} \binom{\mu}{\beta} q'^{c(\beta)} q'^{d(\mu-\beta)} .$$

Because $P_{\alpha a} = \frac{\partial}{\partial q'^{(\alpha)a}} T^{(M)}$ there results, by partial differentiation of both members of (24.2) with respect to $q'^{(\alpha)a}$,

$$(24.3)\quad G_{(\alpha+1)a} = \frac{1}{2}\sum_{\mu=0}^{M}\binom{M}{\mu}g_{cd}{}^{(M-\mu)}\sum_{\beta=0}^{\mu}\binom{\mu}{\beta}\left[q'^{c(\beta)}\delta_a^d\,\delta_\alpha^{\mu-\beta}\right.$$

$$\left. + q'^{d(\mu-\beta)}\delta_a^c\,\delta_\alpha^\beta\right] + \left(1-\delta_\alpha^M\right)\binom{M}{\alpha+1}T_{;0a}{}^{(M-\alpha-1)},$$

$\left(1-\delta_\alpha^M\right)$ indicates deletion if $\alpha = M$. We now have

$$(24.4)\qquad I_{\alpha a} = \frac{1}{2}\sum_{\mu=0}^{M}\binom{M}{\mu}g_{ca}{}^{(M-\mu)}\binom{\mu}{\mu-\alpha}q'^{c(\mu-\alpha)}$$

$$+ \frac{1}{2}\sum_{\mu=0}^{M}\binom{M}{\mu}g_{ad}{}^{(M-\mu)}\binom{\mu}{\alpha}q'^{d(\mu-\alpha)}\ .$$

Here, $I_{\alpha a}$ denotes the first term in the right member of (24.3).

By $\binom{\mu}{\mu-\alpha} = \binom{\mu}{\alpha}$ and $g_{ca} = g_{ac}$ we have, for α : 0 to M,

$$(24.5)\qquad I_{\alpha a} = \sum_{\mu=0}^{M}\binom{M}{\mu}\binom{\mu}{\alpha}g_{ab}{}^{(M-\mu)}q'^{b(\mu-\alpha)}\ .$$

Also, $p_a = g_{ab}q'^b$ and $K_{\alpha a} = \binom{M}{\alpha}p_a{}^{(M-\alpha)}$, so that we may write

$$(24.6)\qquad K_{\alpha a} = \binom{M}{\alpha}\left[g_{ab}q'^b\right]^{(M-\alpha)}\ .$$

By the Leibnitz formula (24.6) becomes

$$(24.7) \qquad K_{\alpha a} = \binom{M}{\alpha} \sum_{\mu=0}^{M-\alpha} \binom{M-\alpha}{\mu} g_{ab}{}^{(\mu)} q'^{b(M-\alpha-\mu)} ,$$

and by replacing the index μ of summation by γ according to the substitution $M - \mu = \gamma$ we have

$$(24.8) \qquad K_{\alpha a} = \binom{M}{\alpha} \sum_{\gamma=\alpha}^{M} \binom{M-\alpha}{M-\gamma} g_{ab}{}^{(M-\gamma)} q'^{b(\gamma-\alpha)} .$$

By replacing γ by μ, and by invoking the identity $\binom{M}{\alpha}\binom{M-\alpha}{M-\mu} = \binom{M}{\mu}\binom{\mu}{\alpha}$ we have from (24.8) for α : 0 to M the result

$$(24.9) \qquad K_{\alpha a} = \sum_{\mu=0}^{M} \binom{M}{\mu}\binom{\mu}{\alpha} g_{ab}{}^{(M-\mu)} q'^{b(\mu-\alpha)} ,$$

since $\binom{\mu}{\alpha} = 0$ for $\mu < \alpha$. From equations (24.5) and (24.9) we have $I_{\alpha a} = K_{\alpha a}$ and

$$G_{(\alpha+1)a} - K_{\alpha a} = \left(1-\delta^{M}_{\alpha}\right)\binom{M}{\alpha+1} T_{;0a}{}^{(M-\alpha-1)} .$$

We now introduce $\overset{+}{K}_{\alpha a}$, the Kawaguchi extensor based on p_a and of range α : 0 to M + 1, and note that $\overset{+}{K}_{(\alpha+1)a}$ and $K_{\alpha a}$ for α : 0 to M are related as follows:

$$K^{+}_{(\alpha+1)a} = \binom{M+1}{\alpha+1} p_a^{(M-\alpha)} = \binom{M+1}{\alpha+1}\binom{M}{\alpha}^{-1} K_{\alpha a} = \frac{M+1}{\alpha+1} K_{\alpha a} .$$

Hence, we may write $K_{\alpha a} = K^{+}_{(\alpha+1)a} - \frac{M-\alpha}{M+1} K^{+}_{(\alpha+1)a}$, and convert the relation $G_{(\alpha+1)a} - K_{\alpha a} = \left(1-\delta^{M}_{\alpha}\right)\binom{M}{\alpha+1} T_{;0a}^{(M-\alpha-1)}$ as follows:

$$G_{(\alpha+1)a} - K^{+}_{(\alpha+1)a} = \frac{M-\alpha}{M+1} K^{+}_{(\alpha+1)a} + \left(1-\delta^{M}_{\alpha}\right)\binom{M}{\alpha+1} T_{;0a}^{(M-\alpha-1)} .$$

We now note that, because the left member of the preceding equation is a reduced range extensor for the range $1 \le \alpha + 1 \le M + 1$, the same may be asserted for the right member. Therefore, the equations

$$\frac{M-\alpha}{M+1} K^{+}_{(\alpha+1)a} - \left(1-\delta^{M}_{\alpha}\right)\binom{M}{\alpha+1} T_{;0a}^{(M-\alpha-1)} = 0$$

constitute an extensor equation. Examination shows that these equations consist of the Euler equations for T, specifically $T_{;0a} - T_{;1a}' = 0$, and derivatives of these equations.

In returning to the construction of primary extensors associated with $\left[h(q,p)\right]^{(M)}$, where $\left[h(q,p)\right]^{(M)} = \left[t(q,p)\right]^{(M)} + \left[V(q)\right]^{(M)}$, we define first the multinomial coefficients $\binom{M}{\alpha,\beta}$ and $\begin{bmatrix}\alpha,\beta\\ M\end{bmatrix}$ for $0 \le \alpha \le M$, $0 \le \beta \le M$, according to

$$(24.10)\qquad \binom{M}{\alpha,\beta} \overset{\text{def}}{\equiv} \begin{cases} \dfrac{M!}{\alpha!\,\beta!\,(M-\alpha-\beta)!} & \text{for } \alpha+\beta \le M \\ 0 & \text{for } \alpha+\beta > M \end{cases} ,$$

i.e., $\binom{M}{\alpha,\beta} = \binom{M}{\alpha}\binom{M-\alpha}{\beta}$, and according to

$$(24.11)\qquad \begin{bmatrix}\alpha,\beta\\ M\end{bmatrix} \overset{\text{def}}{\equiv} \begin{cases} \binom{M}{M-\alpha,M-\beta}\binom{M}{\alpha}^{-1}\binom{M}{\beta}^{-1} & \text{for } \alpha+\beta \ge M \\ 0 & \text{for } \alpha+\beta < M \end{cases} ,$$

i.e., $\begin{bmatrix}\alpha,\beta\\ M\end{bmatrix} = \binom{\alpha}{M-\beta}\binom{M}{\beta}^{-1}$. Next, we define the interchange extensors $g_{\alpha a\cdot\beta b}$ and $g^{\alpha a\cdot\beta b}$ (see [14]) for $0 \le \alpha \le M$, $0 \le \beta \le M$, according to

$$(24.12)\qquad g_{\alpha a\cdot\beta b} \overset{\text{def}}{\equiv} \begin{cases} \binom{M}{\alpha,\beta}\, g_{a\,b}{}^{(M-\alpha-\beta)} & \text{for } \alpha+\beta \le M \\ 0 & \text{for } \alpha+\beta > M \end{cases}$$

and

$$(24.13)\qquad g^{\alpha a\cdot\beta b} \overset{\text{def}}{\equiv} \begin{cases} \begin{bmatrix}\alpha,\beta\\ M\end{bmatrix} g^{ab(\alpha+\beta-M)} & \text{for } \alpha+\beta \ge M \\ 0 & \text{for } \alpha+\beta < M \end{cases} .$$

Now $t(q,p) = \frac{1}{2}\, g^{cd}p_c p_d$, so that by the Leibnitz differentiation formula we have

$$(24.14) \qquad t^{(M)} = \frac{1}{2} \sum_{\mu=0}^{M} \binom{M}{\mu} g^{cd(M-\mu)} \left[p_c p_d\right]^{(\mu)} .$$

By another application of the Leibnitz formula we obtain

$$(24.15) \qquad t^{(M)} = \frac{1}{2} \sum_{\mu=0}^{M} \binom{M}{\mu} g^{cd(M-\mu)} \sum_{\beta=0}^{\mu} \binom{\mu}{\beta} p_c{}^{(\beta)} p_d{}^{(\mu-\beta)} .$$

We now interchange the order of summations in (24.15) to obtain

$$(24.16) \qquad t^{(M)} = \frac{1}{2} \sum_{\beta=0}^{M} \sum_{\mu=\beta}^{M} \binom{M}{\mu}\binom{\mu}{\beta} g^{cd(M-\mu)} p_c{}^{(\beta)} p_d{}^{(\mu-\beta)} .$$

Now replace the index μ of summation by γ according to the substitution

$\mu = \beta + \gamma$ to obtain

$$(24.17) \qquad t^{(M)} = \frac{1}{2} \sum_{\beta=0}^{M} \sum_{\gamma=0}^{M-\beta} \binom{M}{\beta+\gamma}\binom{\beta+\gamma}{\beta} g^{cd(M-\beta-\gamma)} p_c{}^{(\beta)} p_d{}^{(\gamma)}$$

Now substitute $\binom{M}{M-\beta-\gamma,\ \beta}$ for $\binom{M}{\beta+\gamma}\binom{\beta+\gamma}{\beta}$, and substitute $\binom{M}{\beta}^{-1} P_{(M-\beta)c}$ for $p_c{}^{(\beta)}$ and $\binom{M}{\gamma}^{-1} P_{(M-\gamma)d}$ for $p_d{}^{(\gamma)}$. We have

$$(24.18) \qquad t^{(M)} = \frac{1}{2} \sum_{\beta=0}^{M} \sum_{\gamma=0}^{M-\beta} \binom{M}{M-\beta-\gamma,\ \beta} g^{cd(M-\beta-\gamma)} \binom{M}{\beta}^{-1} P_{(M-\beta)c}$$

$$\cdot \binom{M}{\gamma}^{-1} P_{(M-\gamma)d} \quad .$$

By the substitution $\epsilon = M - \gamma$ we replace the index γ of summation by ϵ to obtain

$$(24.19) \qquad t^{(M)} = \frac{1}{2} \sum_{\beta=0}^{M} \sum_{\epsilon=\beta}^{M} \binom{M}{\epsilon-\beta, \beta} g^{cd(\epsilon-\beta)} \binom{M}{\beta}^{-1} P_{(M-\beta)c} \binom{M}{M-\epsilon}^{-1} P_{\epsilon d} \quad .$$

By the substitution $\delta = M - \beta$ we replace the index β of summation by δ to obtain

$$(24.20) \qquad t^{(M)} = \frac{1}{2} \sum_{\delta=0}^{M} \sum_{\epsilon=M-\delta}^{M} \binom{M}{\epsilon-M+\delta, M-\delta} g^{cd(\epsilon+\delta-M)}$$

$$\cdot \binom{M}{M-\delta}^{-1} \binom{M}{M-\epsilon}^{-1} P_{\delta c} P_{\epsilon d} \quad .$$

By $\binom{M}{\epsilon-M+\delta, M-\delta} = \binom{M}{M-\epsilon, M-\delta}$, $\binom{M}{M-\delta} = \binom{M}{\delta}$, and $\binom{M}{M-\epsilon} = \binom{M}{\epsilon}$ we may write (24.20) in the form

$$(24.21) \qquad t^{(M)} = \frac{1}{2} \sum_{\delta=0}^{M} \sum_{\epsilon=M-\delta}^{M} \binom{M}{M-\epsilon, M-\delta} \binom{M}{\delta}^{-1} \binom{M}{\epsilon}^{-1}$$

$$\cdot \, g^{cd(\epsilon+\delta-M)} P_{\delta c} P_{\epsilon d} \quad .$$

By $\begin{bmatrix} \delta, \epsilon \\ M \end{bmatrix} = \binom{M}{M-\epsilon, M-\delta} \binom{M}{\delta}^{-1} \binom{M}{\epsilon}^{-1}$ for $\delta + \epsilon \geq M$, we have from (24.21) the result

$$(24.22)\qquad t^{(M)} = \frac{1}{2}\sum_{\delta=0}^{M}\sum_{\epsilon=M-\delta}^{M}\begin{bmatrix}\delta,\epsilon\\ M\end{bmatrix} g^{cd(\delta+\epsilon-M)}P_{\delta c}P_{\epsilon d}\quad .$$

Now by (24.13) we have $\begin{bmatrix}\delta,\epsilon\\ M\end{bmatrix} g^{cd(\delta+\epsilon-M)} = g^{\delta c\cdot\epsilon d}$ for $\delta + \epsilon \geq M$. Accordingly, (24.22) yields

$$(24.23)\qquad t^{(M)} = \frac{1}{2}\sum_{\delta=0}^{M}\sum_{\epsilon=M-\delta}^{M} g^{\delta c\cdot\epsilon d}P_{\delta c}P_{\epsilon d}\quad .$$

Because $g^{\delta c\cdot\epsilon d} \equiv 0$ for $\delta + \epsilon < M$, the lower limit on the ϵ-summation in (24.23) may be reduced from $M - \delta$ to 0. Upon replacing the indices δ, ϵ, c, d by α, β, a, b, respectively, we obtain from (24.23)

$$(24.24)\qquad t^{(M)} = \frac{1}{2}\sum_{\alpha=0}^{M}\sum_{\beta=0}^{M} g^{\alpha a\cdot\beta b}P_{\alpha a}P_{\beta b}\quad .$$

Accordingly, the differentiated total energy function $\left[h(q,p)\right]^{(M)}$, i.e., $t^{(M)} + V^{(M)}$, becomes identical to the function $H(q,q', \ldots, q^{(M)}, P_{\theta a})$ of the quantities $q,q', \ldots, q^{(M)}$, and of the absolute excovariant extensor $P_{\alpha a}$ of range α : 0 to M, where $H(q,q', \ldots, q^{(M)}, P_{\theta a})$ is given by

$$(24.25)\qquad H(q,q', \ldots ,q^{(M)},P_{\theta a}) = \frac{1}{2}\sum_{\alpha=0}^{M}\sum_{\beta=0}^{M} g^{\alpha a\cdot\beta b}P_{\alpha a}P_{\beta b} + V^{(M)}\ .$$

We now wish to investigate the equality of the primary extensors associated with $H(q,q',\ldots,q^{(M)},P_{\theta a})\left(\equiv\left[h(q,p)\right]^{(M)}\right)$. These primary extensors are defined by referring to Definition (21.1) for the primary extensors $F_{\alpha a}$ and $S_{\alpha a}$ of range α : 0 to M+1 associated with the invariant function $f(x,x',\ldots,x^{(M)},y_{\theta a})$ and a curve C along which $\frac{\partial f}{\partial y_{\alpha a}} = x^{'(\alpha)a}$ for α : 0 to M. The presumption $\frac{\partial f}{\partial y_{\alpha a}} = x^{'(\alpha)a}$ is replaced by

$$(24.26) \qquad \frac{\partial H}{\partial P_{\alpha a}} = q^{'(\alpha)a}$$

for α : 0 to M. Equation (24.26) is shown to be identically true for all trajectories C in our dynamical space as follows: First, differentiate (24.25) to obtain $\frac{\partial H}{\partial P_{\alpha a}} = \sum_{\beta=0}^{M} g^{\alpha a\cdot\beta b}P_{\beta b}$. Then observe that $P_{\beta b} = \binom{M}{\beta}p_b^{(M-\beta)}$ and that $p_b = g_{bd}q^{'d}$ to obtain by the Leibnitz formula from $P_{\beta b} = \binom{M}{\beta}\left[g_{bd}q^{'d}\right]^{(M-\beta)}$ the result $P_{\beta b} = \binom{M}{\beta}\sum_{\mu=0}^{M-\beta}\binom{M-\beta}{\mu}g_{bd}^{(M-\beta-\mu)}q^{'d(\mu)}$. Next, by $\binom{M-\beta}{\mu}g_{bd}^{(M-\beta-\mu)} = \binom{M}{\beta}^{-1}g_{\beta b\cdot\mu d}$, we obtain $P_{\beta b} = \sum_{\mu=0}^{M-\beta} g_{\beta b\cdot\mu d}q^{'d(\mu)}$. Then, substitute this result into $\frac{\partial H}{\partial P_{\alpha a}} = \sum_{\beta=0}^{M} g^{\alpha a\cdot\beta b}P_{\beta b}$ to obtain $\frac{\partial H}{\partial P_{\alpha a}} = \sum_{\beta=0}^{M} g^{\alpha a\cdot\beta b}\sum_{\mu=0}^{M-\beta} g_{\beta b\cdot\mu d}q^{'d(\mu)}$. Next, notice that $g_{\beta b\cdot\mu d} \equiv 0$ if

$\beta + \mu > M$, so that the upper limit on the μ-summation can be raised to M. Then, by reversing the orders of summation on β and μ, obtain

$$\frac{\partial H}{\partial P_{\alpha a}} = \sum_{\mu=0}^{M} \sum_{\beta=0}^{M} g^{\alpha a \cdot \beta b} g_{\beta b \cdot \mu d} q'^{d(\mu)} = \sum_{\mu=0}^{M} \delta^{\alpha}_{\mu} \delta^{a}_{d} q'^{d(\mu)} = q'^{(\alpha)a},$$

independently of the particular trajectory C at hand.

Furthermore, (24.26) is equivalent to a first Hamiltonian equation of motion, namely

$$(24.27) \qquad q'^{a} = \frac{\partial h}{\partial p_a} \quad ,$$

because $\frac{\partial H}{\partial P_{\alpha a}} = q'^{(\alpha)a} = \left[q'^{a}\right]^{(\alpha)} = \left[g^{ad} p_d\right]^{(\alpha)} = \left[\frac{\partial}{\partial p_a}\left(\frac{1}{2} g^{cd} p_c p_d + V(q)\right)\right]^{(\alpha)}$ $= \left[\frac{\partial h}{\partial p_a}\right]^{(\alpha)}$ for α : 0 to M reduces to (24.27) for the case $\alpha = 0$ and to derivatives of (24.27) for the cases $\alpha > 0$.

Accordingly, by Definition (21.1), the primary extensors $E_{\alpha a}$ and $W_{\alpha a}$ associated with $H(q, q', \ldots, q^{(M)}\ P_{\theta a})$ (identical to the differentiated total energy function $\left[h(q,p)\right]^{(M)}$) are defined for all trajectories C by

$$(24.28) \quad \begin{cases} E_{0a} = -\left.\frac{\partial H}{\partial q^{a}}\right|_{P}, \quad E_{\alpha a} = -\left.\frac{\partial H}{\partial q^{(\alpha)a}}\right|_{P} + P_{(\alpha-1)a} \quad \text{for } 1 \le \alpha \le M, \\ \\ E_{(M+1)a} = P_{Ma} \qquad ; \\ \\ W_{\alpha a} = \binom{M+1}{\alpha} E_{(M+1)a}^{\ (M+1-\alpha)} \end{cases}$$

for α : 0 to M+1. By construction, both $E_{\alpha a}$ and $W_{\alpha a}$ are excovariant extensors of range α : 0 to M+1. Furthermore, by (24.25), we have that (24.28) becomes

$$
(24.29)\quad
\begin{aligned}
E_{0a} &= -\frac{1}{2}\sum_{\gamma=0}^{M}\sum_{\beta=0}^{M} g^{\gamma c\cdot\beta b}{}_{;0a}P_{\gamma c}P_{\beta b} - V^{(M)}{}_{;0a} \quad ,\\
E_{\alpha a} &= -\frac{1}{2}\sum_{\gamma=0}^{M}\sum_{\beta=0}^{M} g^{\gamma c\cdot\beta b}{}_{;\alpha a}P_{\gamma c}P_{\beta b} - V^{(M)}{}_{;\alpha a} + P_{(\alpha-1)a}\\
&\qquad \text{for } 1 \le \alpha \le M \quad ,\\
E_{(M+1)a} &= P_{Ma} \quad ;\\
W_{\alpha a} &= \binom{M+1}{\alpha}P_{Ma}{}^{(M+1-\alpha)} \quad .
\end{aligned}
$$

Now $P_{\alpha a} = \binom{M}{\alpha}p_a{}^{(M-\alpha)}$ for α : 0 to M. Accordingly, $P_{(\alpha-1)a} = \binom{M}{\alpha-1}p_a^{(M-\alpha+1)}$ for $1 \le \alpha \le M$, and $P_{Ma} = p_a$. The equations (24.29) for the primary extensors $E_{\alpha a}$ and $W_{\alpha a}$ associated with H become

$$
(24.30)\quad
\left\{
\begin{aligned}
E_{0a} &= -\frac{1}{2}\sum_{\gamma=0}^{M}\sum_{\beta=0}^{M} g^{\gamma c\cdot\beta b}{}_{;0a}P_{\gamma c}P_{\beta b} - V^{(M)}{}_{;0a} \, ,\\
E_{\alpha a} &= -\frac{1}{2}\sum_{\gamma=0}^{M}\sum_{\beta=0}^{M} g^{\gamma c\cdot\beta b}{}_{;\alpha a}P_{\gamma c}P_{\beta b} - V^{(M)}{}_{;\alpha a} + \binom{M}{\alpha-1}p_a{}^{(M-\alpha+1)}\\
&\qquad \text{for } 1 \le \alpha \le M \quad ,\\
E_{(M+1)a} &= p_a \quad ;\\
W_{\alpha a} &= \binom{M+1}{\alpha}p_a{}^{(M+1-\alpha)} \quad ,
\end{aligned}
\right.
$$

for α : 0 to M+1.

We now use (24.30) to examine the extensor equation $E_{\alpha a} = W_{\alpha a}$. We see that rank M+1 of this equation vanishes identically. Rank M of $E_{\alpha a} = W_{\alpha a}$ expresses the equation

$$(24.31) \qquad -\frac{1}{2}\sum_{\gamma=0}^{M}\sum_{\beta=0}^{M} g^{\gamma c\cdot\beta b}{}_{;Ma}P_{\gamma c}P_{\beta b} - V^{(M)}{}_{;Ma} + \binom{M}{M-1}p_a{}' = \binom{M+1}{M}p_a{}' ,$$

which, by $\left[\binom{M+1}{M} - \binom{M}{M-1}\right]p_a{}' = p_a{}'$, is equivalent to

$$(24.32) \qquad -\frac{1}{2}\sum_{\gamma=0}^{M}\sum_{\beta=0}^{M} g^{\gamma c\cdot\beta b}{}_{;Ma}P_{\gamma c}P_{\beta b} - V^{(M)}{}_{;Ma} = p_a{}' .$$

Now for $\gamma + \beta < M$ we have $g^{\gamma c\cdot\beta b}{}_{;Ma} = 0$. For $\gamma + \beta \geq M$ we have $g^{\gamma c\cdot\beta b}{}_{;Ma} = \begin{bmatrix}\gamma,\beta\\ M\end{bmatrix} g^{cb(\gamma+\beta-M)}{}_{;Ma}$. If $\gamma + \beta = M$, the right member of this last equality vanishes. If $\gamma + \beta > M$, $g^{cb(\gamma+\beta-M)}{}_{;Ma} = \left[g^{cb}{}_{;0d}q'^d\right]^{(\gamma+\beta-M-1)}{}_{;Ma}$ $= \left[\sum_{\mu=0}^{\gamma+\beta-M-1}\binom{\gamma+\beta-M-1}{\mu} g^{cb}{}_{;0d}{}^{(\gamma+\beta-M-1-\mu)} q'^{d(\mu)}\right]_{;Ma}$, which vanishes unless $\gamma = \beta = M$, in which case it becomes $g^{cb}{}_{;0a}$. Therefore, (24.32) becomes

$$(24.33) \qquad -\frac{1}{2} g^{cb}{}_{;0a}P_{Mc}P_{Mb} - V^{(M)}{}_{;Ma} = p_a{}' .$$

But $P_{Mc} = p_c$, $P_{Mb} = p_b$, $V^{(M)} = \left[V'\right]^{(M-1)} = \left[V_{;0d}q'^{d}\right]^{(M-1)}$

$= \sum_{\mu=0}^{M-2} \binom{M-1}{\mu} V_{;0d}{}^{(M-1-\mu)} q'^{d(\mu)} + V_{;0d}q'^{d(M-1)}$, and so $V^{(M)}{}_{;Ma} = V_{;0a}$.

Accordingly, (24.33) may now be written as

$$(24.34) \qquad -\tfrac{1}{2}\, g^{cb}{}_{;0a} p_c\, p_b - V_{;0a} = p_a{}' .$$

But, because $h = \tfrac{1}{2} g^{cb} p_c p_b + V(q)$, we have $\left.\dfrac{\partial h}{\partial q^a}\right|_p = \tfrac{1}{2}\, g^{cb}{}_{;0a} p_c p_b + V_{;0a}$.

Thus rank M of $E_{\alpha a} = W_{\alpha a}$, expressed as (23.34), is equivalent to the second Hamiltonian equation of motion

$$(24.35) \qquad \left.\frac{\partial h}{\partial q^a}\right|_p = -\, p_a{}' .$$

We now consider $E_{\alpha a} = W_{\alpha a}$ for ranks α such that $0 \le \alpha \le M$. By use of $\left[\binom{M+1}{\alpha} - \binom{M}{\alpha-1}\right] p_a{}^{(M+1-\alpha)} = \binom{M}{\alpha} p_a{}^{(M+1-\alpha)}$ whenever $1 \le \alpha \le M$, and by use of $\binom{M+1}{\alpha} = \binom{M}{\alpha}(=1)$ when $\alpha = 0$, we have that $E_{\alpha a} = W_{\alpha a}$ expresses

$$(24.36) \qquad -\tfrac{1}{2} \sum_{\gamma=0}^{M} \sum_{\beta=0}^{M} g^{\gamma c\cdot\beta b}{}_{;\alpha a} P_{\gamma c} P_{\beta b} - V^{(M)}{}_{;\alpha a} = \binom{M}{\alpha} p_a{}^{(M+1-\alpha)}$$

for $0 \le \alpha \le M$. Now $P_{\gamma c} = \binom{M}{\gamma} p_c{}^{(M-\gamma)}$, $P_{\beta b} = \binom{M}{\beta} p_b{}^{(M-\beta)}$, and $g^{\gamma c\cdot\beta b} = \begin{bmatrix}\gamma,\beta\\ M\end{bmatrix} g^{cb(\gamma+\beta-M)}$ for $\gamma + \beta \ge M$; $g^{\gamma c\cdot\beta b} = 0$ for $\gamma + \beta < M$. Because

$g^{\gamma c\cdot\beta b} = 0$ for $\gamma + \beta < M$, the lower limit on the β-summation in the left member of (24.36) can be raised, for each γ, to $M - \gamma$, after which $g^{\gamma c\cdot\beta b}$ can be replaced by $\begin{bmatrix}\gamma,\beta\\ M\end{bmatrix} g^{cb(\gamma+\beta-M)}$ in (24.36). With the indicated replacements for $P_{\gamma c}$ and $P_{\beta b}$, we have that (24.36) becomes

$$(24.37)\quad -\frac{1}{2}\sum_{\gamma=0}^{M}\sum_{\beta=M-\gamma}^{M}\binom{M}{\beta}\binom{M}{\gamma}\begin{bmatrix}\gamma,\beta\\ M\end{bmatrix} g^{cb(\gamma+\beta-M)}{}_{;\alpha a}\,p_c{}^{(M-\gamma)}p_b{}^{(M-\beta)} - V^{(M)}{}_{;\alpha a}$$

$$= \binom{M}{\alpha} p_a{}^{(M+1-\alpha)} .$$

Notice now that $\begin{bmatrix}\gamma,\beta\\ M\end{bmatrix} = \binom{M}{M-\gamma, M-\beta}\binom{M}{\gamma}^{-1}\binom{M}{\beta}^{-1}$, so that $\binom{M}{\gamma}\binom{M}{\beta}\begin{bmatrix}\gamma,\beta\\ M\end{bmatrix}$ $= \binom{M}{M-\gamma, M-\beta} = \binom{M}{M-\gamma}\binom{\gamma}{M-\beta}$, and (24.37) becomes

$$(24.38)\quad -\frac{1}{2}\sum_{\gamma=0}^{M}\sum_{\beta=M-\gamma}^{M}\binom{M}{M-\gamma}\binom{\gamma}{M-\beta} g^{cb(\gamma+\beta-M)}{}_{;\alpha a}\,p_c{}^{(M-\gamma)}p_b{}^{(M-\beta)} - V^{(M)}{}_{;\alpha a}$$

$$= \binom{M}{\alpha} p_a{}^{(M+1-\alpha)} .$$

Equation (24.38) may now be altered by the replacement $g^{cb(\gamma+\beta-M)}{}_{;\alpha a} = \binom{\gamma+\beta-M}{\alpha} g^{cb}{}_{;0a}{}^{(\gamma+\beta-M-\alpha)}$, where $g^{cb}{}_{;0a}{}^{(\gamma+\beta-M-\alpha)} \overset{\text{def}}{\equiv} 0$ if $\alpha > \gamma + \beta - M$. We obtain

$$(24.39)\quad -\frac{1}{2}\sum_{\gamma=0}^{M}\sum_{\beta=M-\gamma}^{M}\binom{M}{M-\gamma}\binom{\gamma}{M-\beta}\binom{\gamma+\beta-M}{\alpha} g^{cb}{}_{;0a}{}^{(\gamma+\beta-M-\alpha)}p_c{}^{(M-\gamma)}p_b{}^{(M-\beta)}$$

$$- V^{(M)}{}_{;\alpha a} = \binom{M}{\alpha} p_a{}^{(M+1-\alpha)} .$$

We now notice that, because $\binom{\gamma+\beta-M}{\alpha} = 0$ for $\beta < M - \gamma + \alpha$, the lower limit on the β-summation in the left member of (24.39) can be raised from $M - \gamma$ to $M - \gamma + \alpha$. We next introduce the index μ of summation in place of β in (24.39) according to the substitution $\mu = \gamma + \beta - M - \alpha$. We get

$$(24.40) \quad -\frac{1}{2} \sum_{\gamma=0}^{M} \sum_{\mu=0}^{\gamma-\alpha} \binom{M}{M-\gamma}\binom{\gamma}{\gamma-\alpha-\mu}\binom{\mu+\alpha}{\alpha} g^{cb}{}_{;0a}{}^{(\mu)}p_c{}^{(M-\gamma)}p_b{}^{(\gamma-\alpha-\mu)} - v^{(M)}{}_{;\alpha a}$$

$$= \binom{M}{\alpha} p_a{}^{(M+1-\alpha)} \quad .$$

We now replace the index γ of summation in (24.40) by δ according to the replacement $M - \gamma = \delta$. We get

$$(24.41) \quad -\frac{1}{2} \sum_{\delta=0}^{M} \sum_{\mu=0}^{M-\delta-\alpha} \binom{M}{\delta}\binom{M-\delta}{M-\delta-\alpha-\mu}\binom{\mu+\alpha}{\alpha} g^{cb}{}_{;0a}{}^{(\mu)}p_c{}^{(\delta)}p_b{}^{(M-\delta-\alpha-\mu)}$$

$$- v^{(M)}{}_{;\alpha a} = \binom{M}{\alpha} p_a{}^{(M+1-\alpha)} \quad .$$

Now we observe that $\binom{M}{\delta}\binom{M-\delta}{M-\delta-\alpha-\mu}\binom{\mu+\alpha}{\alpha}$ is equal to $\binom{M}{\alpha}\binom{M-\alpha}{\mu}\binom{M-\alpha-\mu}{\delta}$.

We make this substitution into (24.41) while reversing the order of the δ- and μ-summations, with appropriate alteration of the upper and lower limits to get

$$(24.42) \quad -\frac{1}{2}\sum_{\mu=0}^{M-\alpha}\sum_{\delta=0}^{M-\alpha-\mu}\binom{M}{\alpha}\binom{M-\alpha}{\mu}\binom{M-\alpha-\mu}{\delta}g^{cb}{}_{;0a}{}^{(\mu)}p_c{}^{(\delta)}p_b{}^{(M-\delta-\alpha-\mu)}$$

$$- V^{(M)}{}_{;\alpha a} = \binom{M}{\alpha}p_a{}^{(M+1-\alpha)}.$$

At this point we replace $\sum_{\delta=0}^{M-\alpha-\mu}\binom{M-\alpha-\mu}{\delta}p_c{}^{(\delta)}p_b{}^{(M-\delta-\alpha-\mu)}$ in (24.42) by $(p_cp_b)^{(M-\alpha-\mu)}$, the equivalence following by Leibnitz's formula. Equation (24.42) becomes

$$(24.43) \quad -\frac{1}{2}\binom{M}{\alpha}\sum_{\mu=0}^{M-\alpha}\binom{M-\alpha}{\mu}g^{cb}{}_{;0a}{}^{(\mu)}(p_cp_b)^{(M-\alpha-\mu)} - V^{(M)}{}_{;\alpha a}$$

$$= \binom{M}{\alpha}p_a{}^{(M+1-\alpha)} \quad .$$

By use of Leibnitz's formula we may replace $\sum_{\mu=0}^{M-\alpha}\binom{M-\alpha}{\mu}g^{cb}{}_{;0a}{}^{(\mu)}(p_cp_b)^{(M-\alpha-\mu)}$ in (24.43) by $\left[g^{cb}{}_{;0a}p_cp_b\right]^{(M-\alpha)}$ to obtain

$$(24.44) \quad -\frac{1}{2}\binom{M}{\alpha}\left[g^{cb}{}_{;0a}p_cp_b\right]^{(M-\alpha)} - V^{(M)}{}_{;\alpha a} = \binom{M}{\alpha}p_a{}^{(M+1-\alpha)} \quad .$$

Next, we substitute $\binom{M}{\alpha}V_{;0a}{}^{(M-\alpha)}$ for $V^{(M)}{}_{;\alpha a}$ in (24.44), which may then be rewritten in the form

$$(24.45) \quad -\binom{M}{\alpha}\left[\frac{1}{2}g^{cb}{}_{;0a}p_cp_b + V_{;0a}\right]^{(M-\alpha)} = \binom{M}{\alpha}\left[p_a{}'\right]^{(M-\alpha)} \quad .$$

But $h(q,p) = \frac{1}{2} g^{cb} p_c p_b + V(q)$, so that $\left.\frac{\partial h}{\partial q^a}\right|_p = \frac{1}{2} g^{cb}{}_{;0a} p_c p_b + V_{;0a}$.

Now (24.45) yields

$$(24.46) \qquad -\binom{M}{\alpha}\left[\left.\frac{\partial h}{\partial q^a}\right|_p + p_a{}'\right]^{(M-\alpha)} = 0 .$$

Thus (24.46), which is equivalent to the equality of the primary extensors $E_{\alpha a}$ and $W_{\alpha a}$ associated with $H(q,q', \ldots, q^{(M)}, P_{\theta a})$ $\left(\equiv \left[h(q,p)\right]^{(M)}\right)$ whenever $0 \leq \alpha \leq M$, expresses the $(M-\alpha)$th derivative of the second Hamiltonian equation of motion (see (24.35)). Rank M of $E_{\alpha a} = W_{\alpha a}$ is precisely equivalent to the Hamiltonian equation (24.35), as noted previously. We also noted that rank $M+1$ of $E_{\alpha a} = W_{\alpha a}$ vanished identically. Therefore, we have completed the proof of the assertion that not only is the first Hamiltonian equation, namely $q'^a = \left.\frac{\partial h}{\partial p_a}\right|_q$, equivalent to the basic assumption which permits the construction of the primary extensors $E_{\alpha a}$ and $W_{\alpha a}$ associated with $H(q,q', \ldots, q^{(M)}, P_{\theta a})\left(=\left[h(q,p)\right]^{(M)}\right)$, and not only is this equation identically satisfied along all trajectories in our dynamical space, but also the equality of the primary extensors $E_{\alpha a}$ and $W_{\alpha a}$ of range α : 0 to $M+1$ (a) expresses the second Hamiltonian equation $\left.\frac{\partial h}{\partial q^a}\right|_p = -p_a{}'$ through rank M, the highest nonvanishing rank of $E_{\alpha a} - W_{\alpha a}$, and (b) expresses the $(M-\alpha)$th derivative of the second Hamiltonian equation through rank α for each α such that $0 \leq \alpha \leq M$.

We now consider briefly some variational aspects of our dynamical formulation. First of all, we note that the Hamiltonian equation of motion $\left.\frac{\partial h}{\partial q^a}\right|_p = -p_a{}'$ is obtainable as the Euler (or Hamiltonian) equations associated with the simple extensor-generalized calculus of variations problem based upon $\int_{t_1}^{t_2} \sum_{\alpha=0}^{1} A_{\alpha a} V^{a(\alpha)} dt$, where $A_{0a} = -\left.\frac{\partial h}{\partial q^a}\right|_p$, $A_{1a} = p_a$, and where $A_{\alpha a}$, with α : 0 to 1, is the first primary extensor associated with $h(q,p)$, i.e., $\sum_{\alpha=0}^{1} (-1)^\alpha A_{\alpha a}{}^{(\alpha)} = 0$ is equivalent to $\left.\frac{\partial h}{\partial q^a}\right|_p + p_a{}' = 0.$

Next, we note that the Euler (or Hamiltonian) equations associated with the simple extensor-generalized calculus of variations problem based upon $\int_{t_1}^{t_2} \sum_{\alpha=0}^{M+1} E_{\alpha a} V^{a(\alpha)} dt$ vanish identically. For we have

$$\sum_{\alpha=0}^{M+1} (-1)^\alpha E_{\alpha a}{}^{(\alpha)} = E_{0a} + \sum_{\alpha=1}^{M} (-1)^\alpha E_{\alpha a}{}^{(\alpha)} + (-1)^{M+1} E_{(M+1)a}{}^{(M+1)}$$

$$= -\frac{1}{2} \sum_{\gamma=0}^{M} \sum_{\beta=0}^{M} g^{\gamma c \cdot \beta b}{}_{;0a} P_{\gamma c} P_{\beta b} - V^{(M)}{}_{;0a} - \frac{1}{2} \sum_{\alpha=1}^{M} (-1)^\alpha \sum_{\gamma=0}^{M} \sum_{\beta=0}^{M} \left[g^{\gamma c \cdot \beta b}{}_{;\alpha a} P_{\gamma c} P_{\beta b} \right]^{(\alpha)}$$

$$- \sum_{\alpha=1}^{M} (-1)^\alpha \left[V^{(M)}{}_{;\alpha a} \right]^{(\alpha)} + \sum_{\alpha=1}^{M} (-1)^\alpha \binom{M}{\alpha-1} p_a{}^{(M+1)} + (-1)^{M+1} p_a{}^{(M+1)}.$$

By combining terms, this simplifies to $\sum_{\alpha=0}^{M+1} (-1)^{\alpha} E_{\alpha a}{}^{(\alpha)}$

$$= -\frac{1}{2} \sum_{\alpha=0}^{M} (-1)^{\alpha} \sum_{\gamma=0}^{M} \sum_{\beta=0}^{M} \left[g^{\gamma c \cdot \beta b}{}_{;\alpha a} P_{\gamma c} P_{\beta b} \right]^{(\alpha)} - \sum_{\alpha=0}^{M} (-1)^{\alpha} \left[V^{(M)}{}_{;\alpha a} \right]^{(\alpha)}$$

$$+ \left[\sum_{\alpha=1}^{M+1} (-1)^{\alpha} \binom{M}{\alpha-1} \right] P_a{}^{(M+1)}.$$

Now by reference to equations (24.36) through (24.44) we see that $\sum_{\gamma=0}^{M} \sum_{\beta=0}^{M} g^{\gamma c \cdot \beta b}{}_{;\alpha a} P_{\gamma c} P_{\beta b}$ may be replaced by $\binom{M}{\alpha} \left[g^{cb}{}_{;0a} p_c p_b \right]^{(M-\alpha)}$. We also recall that $V^{(M)}{}_{;\alpha a}$ is equal to $\binom{M}{\alpha} V_{;0a}{}^{(M-\alpha)}$. Therefore, we now have $\sum_{\alpha=0}^{M+1} (-1)^{\alpha} E_{\alpha a}{}^{(\alpha)} = -\frac{1}{2} \left[\sum_{\alpha=0}^{M} (-1)^{\alpha} \binom{M}{\alpha} \right] \left[g^{cb}{}_{;0a} p_c p_b \right]^{(M)}$ $- \left[\sum_{\alpha=0}^{M} (-1)^{\alpha} \binom{M}{\alpha} \right] V_{;0a}{}^{(M)} + \left[\sum_{\alpha=1}^{M+1} (-1)^{\alpha} \binom{M}{\alpha-1} \right] p_a{}^{(M+1)}$. Because for $M > 0$ the expression $\sum_{\alpha=0}^{M} (-1)^{\alpha} \binom{M}{\alpha}$ may be regarded as the binomial expansion for $\left[1 + (-1) \right]^{M}$, we have that it vanishes identically. By the substitution $\beta = \alpha - 1$ we have $\sum_{\alpha=1}^{M+1} (-1)^{\alpha} \binom{M}{\alpha-1} = \sum_{\beta=0}^{M} (-1)^{\beta} \binom{M}{\beta} = 0$. Accordingly, it is indeed the case that $\sum_{\alpha=0}^{M+1} (-1)^{\alpha} E_{\alpha a}{}^{(\alpha)} \equiv 0$.

Finally, we note that, because $W_{\alpha a}$ has a Kawaguchi structure, we have that the Euler (or Hamiltonian) equations associated with the simple extensor-generalized calculus of variations problem $\int_{t_1}^{t_2} \sum_{\alpha=0}^{M+1} W_{\alpha a} V^{a(\alpha)} dt$ vanish identically, i.e., $\sum_{\alpha=0}^{M+1} (-1)^{\alpha} W_{\alpha a}{}^{(\alpha)} \equiv 0$.

CHAPTER IV

THE SET OF FUNDAMENTAL EXTENSORS ASSOCIATED WITH $E_{\alpha a}$

25. Construction of the generalized Kawaguchi extensors $(L,M^*)_{\alpha a}$. As a first step toward the construction of (1) covariant tensors using the excovariant structure of $E_{\alpha a}$, and of (2) a class of variational integral problems generated by $E_{\alpha a}$, let us proceed to devise a method which will ultimately result in a family $\mathcal{S}(E_{\alpha a})$ of fundamental extensors associated with the extensor $E_{\alpha a}$ of range α : 0 to M, $M > 0$.

Now an important property of the Kawaguchi extensor $K_{\alpha a}$ of range α : 0 to M, $M > 0$, $K_{\alpha a} = \binom{M}{\alpha} T_a{}^{(M-\alpha)}$, is that its contraction $\sum_{\alpha=0}^{M} K_{\alpha a} V^{a(\alpha)}$ with an arbitrary extensor of the type $V^{a(\alpha)}$ produces the Mth derivative of the tensor contraction $T_a V^a$. Here, T_a and V^a are functions of a curve parameter t of class C^M. Our first problem is to construct a generalization based on this contraction property by defining a set of quantities $(L,M^*)_{\alpha a}$ of range α : 0 to M^* having the property along any arc $x^a = x^a(t)$ of class C^{M^*} that

$$\sum_{\alpha=0}^{M^*} (L,M^*)_{\alpha a} V^{a(\alpha)} = \left[\sum_{\alpha=L}^{M} \binom{\alpha}{L} E_{\alpha a} V^{a(\alpha-L)} \right]^{(D)}.$$

Here, $E_{\alpha a}$ is an extensor of range α : 0 to M, $M > 0$, L is an integer such that $0 \leq L \leq M$, M^* is a nonnegative integer such that $M^* \geq M - L$, and D is a nonnegative integer, which we will see to be given by $D = M^* - M + L$. The expression $\left[\sum_{\alpha=L}^{M} \binom{\alpha}{L} E_{\alpha a} V^{a(\alpha-L)} \right]^{(D)}$ is a derivative of a reduced range contraction whose essential elements are the extensor $E_{\alpha a}$, the tensor V^a, and the integer L. Accordingly, we adopt the

Definition (25.1). $\mathrm{EVL} = \sum_{\alpha=L}^{M} \binom{\alpha}{L} E_{\alpha a} V^{a(\alpha-L)}$.

Section 11 covering a certain invariant as a reduced range extensor contraction, as well as the presentation in [6], pp. 272-275 of the extensor quotient law, reveals that if $E_{\alpha a}$ is an absolute excovariant extensor of range α : 0 to M, $M > 0$, if $V^a(t)$ is an arbitrary class c^M tensor, and if $0 \leq L \leq M$, then the expression EVL is an invariant, as well as each of its total derivatives with respect to the parameter t. The two-parameter analog is also verified in Section 11. Therefore, if the quantities $(L,M^*)_{\alpha a}$ can be constructed, then because EVL and $(\mathrm{EVL})^{(D)}$ are invariants, it follows by the quotient law for extensors, together with the requirement that $E_{\alpha a}$ be independent of V^a, that the quantities $(L,M^*)_{\alpha a}$ are components of an extensor of range α : 0 to M^*. Because $0 \leq L \leq M$, $L \geq M - M^*$, it follows that, with M^* fixed, the set $\{(L,M^*)_{\alpha a} | 0 \leq L \leq M,\ L \geq M - M^*\}$ constitutes a family of extensors, all of range α : 0 to M^*, which are constructed from the derivatives of the components of the given extensor $E_{\alpha a}$. Thus generalizations of the Kawaguchi extensor are produced.

An investigation of the structure of $(EVL)^{(D)}$, D a nonnegative integer, leads to the consideration of the following tentative definition for $(L,M^*)_{\alpha a}$:

$$(25.1) \qquad (L,M^*)_{\alpha a} = \sum_{\mu=D-\lambda}^{D} \binom{D}{\mu} \binom{\alpha+\mu+L-D}{L} E_{(\alpha+\mu+L-D)a}{}^{(\mu)}$$

for α : 0 to M^*, with $\lambda = \alpha$ for α : 0 to D, and $\lambda = D$ if $\alpha \geq D$. Here, $M^* \geq M - L$, and D is defined by $D = M^* - M + L$. Hence, $M^* - M = D - L$.

Observations: We note that, if we extend the range of $E_{\alpha a}$ by defining $E_{(M+\gamma)a} \equiv 0$ for $\gamma \geq 1$, we may then write $(L,M^*)_{(M^*+\gamma)a} = 0$ for $\gamma \geq 1$, as for each μ such that $D - \lambda \leq \mu \leq D$ we have $E_{(M^*+\gamma+\mu+L-D)a} = E_{(M+\gamma+\mu)a} = 0$.

Also, it follows from (25.1) that in the computation of $(L,M^*)_{0a}$ we have $\lambda = 0$, so that μ is fixed at D. Hence we may write for the derivative $(L,M^*)_{0a}{}'$ the result

$$(25.2) \qquad (L,M^*)_{0a}{}' = E_{La}{}^{(D+1)} \quad .$$

Now $(L,M^*+1)_{\alpha a}$ is given by the right member of (25.1) with D replaced by D + 1. There follows $(L,M^*+1)_{0a} = E_{La}{}^{(D+1)}$, or

$$(25.3) \qquad (L,M^*)_{0a}{}' = (L,M^*+1)_{0a} \; .$$

We next observe, by a product rule for differentiation, that the derivative $\left[\sum_{\alpha=0}^{M^*} (L,M^*)_{\alpha a} v^{(\alpha)a}\right]'$, which will be involved in an induction proof, may be converted according to

$$(25.4) \left[\sum_{\alpha=0}^{M^*} (L,M^*)_{\alpha a} v^{(\alpha)a}\right]' = \sum_{\alpha=0}^{M^*} (L,M^*)_{\alpha a}{}' v^{(\alpha)a} + \sum_{\beta=0}^{M^*} (L,M^*)_{\beta b} v^{(\beta+1)b} \; .$$

By $(L,M^*)_{(M^*+1)a} = 0$, and by an expansion of the right member of (25.4), we obtain

$$(25.5) \left[\sum_{\alpha=0}^{M^*} (L,M^*)_{\alpha a} v^{(\alpha)a}\right]' = (L,M^*)_{0a}{}' v^{a}$$

$$+ \sum_{\alpha=1}^{M^*+1} \left[(L,M^*)_{\alpha a}{}' + (L,M^*)_{(\alpha-1)a}\right] v^{(\alpha)a} \; .$$

Now the expressions in the bracket in the right member of (25.5) may be evaluated by (25.1) as follows: First, note that for α : 1 to M* + 1 we have

$$(25.6)\quad (L,M^*)_{(\alpha-1)a} = \sum_{\mu=D-\nu}^{D} \binom{D}{\mu}\binom{\alpha-1+\mu+L-D}{L} E_{(\alpha-1+\mu+L-D)a}^{(\mu)},$$

with $\nu = \alpha - 1$ if $0 \leq \alpha - 1 \leq D$, $\nu = D$ if $\alpha - 1 \geq D$, for $\alpha - 1$: 0 to M*. We now replace ν with $\lambda - 1$ and obtain

$$(25.7)\quad (L,M^*)_{(\alpha-1)a} = \sum_{\mu=D+1-\lambda}^{D} \binom{D}{\mu}\binom{\alpha-1+\mu+L+D}{L} E_{(\alpha-1+\mu+L+D)a}^{(\mu)} \qquad ,$$

with $\lambda = \alpha$ if $1 \leq \alpha \leq D + 1$, $\lambda = D + 1$ if $\alpha \geq D + 1$, for α : 1 to M* + 1.

Second, with regard to $(L,M^*)_{\alpha a}'$ we have

$$(25.8)\quad (L,M^*)_{\alpha a}' = \sum_{\nu=D-\lambda}^{D} \binom{D}{\nu}\binom{\alpha+\nu+L-D}{L} E_{(\alpha+\nu+L-D)a}^{(\nu+1)} \qquad ,$$

or, by substituting $\nu = \mu - 1$,

$$(25.9)\quad (L,M^*)_{\alpha a}' = \sum_{\mu=D+1-\lambda}^{D+1} \binom{D}{\mu-1}\binom{\alpha+\mu-1+L-D}{L} E_{(\alpha+\mu-1+L-D)a}^{(\mu)} \qquad ,$$

where $\lambda = \alpha$ if $0 \leq \alpha \leq D$, $\lambda = D$ if $\alpha \geq D$, for α : 0 to M*.

Now, if D = 0, we have by (25.9) the result

$$(25.10) \qquad (L,M^*)_{\alpha a}{}' = \binom{\alpha+L}{L} E_{(\alpha+L)a}{}^{\lambda}, \quad D = 0.$$

For $D \geq 0$, the expression $(L,M^*)_{0a}'$ is given by equation (25.2). If $\alpha > 0$ and $D > 0$, we may extract from the summation in (25.9) the term corresponding to $\mu = D + 1$ to obtain

$$(25.11) \quad (L,M^*)_{\alpha a}{}' = \binom{\alpha+L}{L} E_{(\alpha+L)a}{}^{(D+1)} + \sum_{\mu=D+1-\lambda}^{D} \binom{D}{\mu-1} \binom{\alpha+\mu-1+L-D}{L} E_{(\alpha+\mu-1+L-D)a}{}^{(\mu)},$$

where $\lambda = \alpha$ if $1 \leq \alpha \leq D$, $\lambda = D$ if $\alpha \geq D$, for $\alpha : 1$ to M^*.

We now compare (25.7) to (25.11). In (25.7) we have $\lambda = \alpha$ if $1 \leq \alpha \leq D + 1$, $\lambda = D + 1$ if $\alpha \geq D + 1$, for $\alpha : 1$ to $M^* + 1$. We note these differences: (1) In (25.11) the range of α is 1 to M^*, whereas in (25.7) it is 1 to $M^* + 1$. (2) In (25.11) $\lambda = D$ for $\alpha \geq D + 1$, whereas in (25.7) $\lambda = D + 1$ for $\alpha \geq D + 1$. We note, however, that, because $E_{(M^*+1+L)a} = E_{(M+1+D)a} = 0$, and because $E_{(M^*+1+\mu+L-D-1)a} = E_{(M+\mu)a} = 0$ for $1 \leq \mu \leq D$, the range on α in (25.11) can be extended up to $M^* + 1$. Furthermore, (25.11) is needed only for $\alpha : 1$ to $M^* + 1$ in its application to (25.5). Also, if the symbol $\binom{D}{-1}$ is assigned the value zero, then the requirement $\lambda = D$ for $\alpha \geq D$ in (25.11) may be changed to $\lambda = D + 1$ for $\alpha \geq D + 1$. The effect of this change is to convert the summation range for μ in (25.11) from $1 \leq \mu \leq D$ to $0 \leq \mu \leq D$ for $\alpha \geq D + 1$. A term is introduced thereby containing the symbolic factor $\begin{pmatrix} D \\ -1 \end{pmatrix}$. In the following we denote by (25.11a) the augmented form of (25.11) in which α has the range 1 to $M^* + 1$, and $\lambda = D + 1$ for $\alpha \geq D + 1$. With equations (25.7) and (25.11a) as background, we are now ready to consider the following fundamental proposition:

Theorem (25.1). Suppose that (1) $E_{\alpha a}$ is an absolute excovariant extensor of range α : 0 to M, $M > 0$, (2) M^* is a nonnegative integer, (3) L is an integer such that $0 \leq L \leq M$, $L \geq M - M^*$, (4) EVL and $(L,M^*)_{\alpha a}$ are given, respectively, by Definition (25.1) and equation (25.1). It then follows that

$$\sum_{\alpha=0}^{M^*} (L,M^*)_{\alpha a} V^{a(\alpha)} = (EVL)^{(D)} .$$

Remark: It may be noted by (25.1) that $(M,M^*)_{\alpha a}$ has precisely the Kawaguchi structure $\begin{pmatrix} M^* \\ \alpha \end{pmatrix} E_{Ma}{}^{(M^*-\alpha)}$ for α : 0 to M^*, so that, for fixed M^*, the set $\{(L,M^*)_{\alpha a} | 0 \leq L \leq M,\ L \geq M - M^*\}$ includes an extensor, namely $(M,M^*)_{\alpha a}$ with a pure Kawaguchi structure. Here, when L = M, the invariant EVL reduces to the tensor contraction $E_{Ma}V^a$.

Proof of Theorem (25.1): The special case of this theorem for which $M^* = M - L$, and accordingly $D = 0$, may be established from (25.1) as follows: We first note that in this case $\lambda = 0$, and therefore μ is fixed at zero. Accordingly, for $M^* = M - L$, $D = 0$, we have the formula

$$(L,\ M - L)_{\alpha a} = \binom{\alpha+L}{L} E_{(\alpha+L)a} \quad ,$$

and consequently we note that

$$\sum_{\beta=0}^{M-L} (L,M - L)_{\beta a} V^{(\beta)a} = \sum_{\beta=0}^{M-L} \binom{\beta+L}{L} E_{(\beta+L)a} V^{(\beta)a}$$

$$= \sum_{\alpha=L}^{M} \binom{\alpha}{L} E_{\alpha a} V^{(\alpha-L)a} = (EVL)^{(0)} \quad .$$

Next, the special case for which $M^* = M - L + 1$, $D = 1$, is established from (25.1) as follows: We first note that, if $\alpha = 0$, we have $\lambda = 0$, and therefore μ is fixed at $D = 1$. We also note that, if $\alpha > 0$, $\lambda = 1$, and therefore μ is restricted to the set 0,1. We get

$$(L,M-L+1)_{0a} = E_{La}{}' , \quad (L,M-L+1)_{\alpha a} = \binom{\alpha+L-1}{L} E_{(\alpha+L-1)a} + \binom{\alpha+L}{L} E_{(\alpha+L)a}{}'$$

for α : 1 to M*,

and consequently we note that

$$\sum_{\beta=0}^{M-L+1} (L,M-L+1)_{\beta a} V^{(\beta)a} = E_{La}{}' V^{a} + \sum_{\beta=1}^{M-L+1} \binom{\beta+L-1}{L} E_{(\beta+L-1)a} V^{(\beta)a}$$

$$+ \sum_{\beta=1}^{M-L+1} \binom{\beta+L}{L} E_{(\beta+L)a}{}' V^{(\beta)a}$$

$$= E_{La}{}' V^{a} + \sum_{\alpha=L}^{M} \binom{\alpha}{L} E_{\alpha a} V^{a(\alpha-L+1)}$$

$$+ \sum_{\alpha=L+1}^{M+1} \binom{\alpha}{L} E_{\alpha a}{}' V^{a(\alpha-L)} \quad .$$

The term $E_{La}{}' V^{a}$ can be absorbed into the second summation by dropping the lower limit on this summation from L + 1 to L. The upper limit can be lowered from M + 1 to M, as $E_{(M+1)a} \overset{\text{def}}{\equiv} 0$. We have

$$\sum_{\beta=0}^{M-L+1} (L,M-L+1)_{\beta a} V^{(\beta)a} = \sum_{\alpha=L}^{M} \binom{\alpha}{L} E_{\alpha a}{}' V^{a(\alpha-L)} + \sum_{\alpha=L}^{M} \binom{\alpha}{L} E_{\alpha a} V^{a(\alpha-L+1)}$$

$$= \left[\sum_{\alpha=L}^{M} \binom{\alpha}{L} E_{\alpha a} V^{a(\alpha-L)} \right]' = (EVL)' .$$

We now proceed by induction and show that the validity of the theorem for L, M*, D with D $\geq$ 1 implies the validity of the theorem for L, M* + 1, D + 1, assuming the necessary differentiability. As a first step, we note that, by virtue of the induction assumption, by equation (25.3), and by our observation that $(L,M^*)_{(M^*+1)a} = 0$, we may write for $D \geq 1$ the following:

$$(EVL)^{(D+1)} = \left[(EVL)^{(D)}\right]' = \left[\sum_{\alpha=0}^{M^*} (L,M^*)_{\alpha a} V^{(\alpha)a}\right]'$$

$$= \sum_{\beta=0}^{M^*} (L,M^*)_{\beta a}{}' V^{(\beta)a} + \sum_{\beta=0}^{M^*} (L,M^*)_{\beta a} V^{(\beta+1)a}$$

$$= (L,M^*)_{0a}{}' V^{a} + \sum_{\alpha=1}^{M^*} (L,M^*)_{\alpha a}{}' V^{(\alpha)a} + \sum_{\alpha=1}^{M^*+1} (L,M^*)_{(\alpha-1)a} V^{(\alpha)a}$$

$$= (L,M^*+1)_{0a} V^{a} + \sum_{\alpha=1}^{M^*+1} \left[(L,M^*)_{\alpha a}{}' + (L,M^*)_{(\alpha-1)a}\right] V^{(\alpha)a} .$$

From the equations (25.7) and (25.11a) and the formula $\binom{D}{\mu-1} + \binom{D}{\mu} = \binom{D+1}{\mu}$, we have

$$(EVL)^{(D+1)} = (L,M^*+1)_{0a} V^{a} + \sum_{\alpha=1}^{M^*+1} V^{(\alpha)a}\left[\binom{\alpha+L}{L} E_{(\alpha+L)a}{}^{(D+1)} + \sum_{\mu=D+1-\lambda}^{D} \binom{D+1}{\mu}\binom{\alpha+\mu-1+L-D}{L} E_{(\alpha+\mu-1+L-D)a}{}^{(\mu)}\right] ,$$

where $\lambda = \alpha$ if $1 \leq \alpha \leq D + 1$, $\lambda = D + 1$ if $D + 1 \leq \alpha \leq M^* + 1$. But the square bracket is just equal to $(L,M^* + 1)_{\alpha a}$ for each α such that $1 \leq \alpha \leq M^* + 1$. Accordingly, we have

$$(EVL)^{(D+1)} = \sum_{\alpha=0}^{M^*+1} (L, M^* + 1)_{\alpha a} V^{(\alpha)a} \quad .$$

The induction is completed, and the theorem is proved.

Comment: Given $M^* \geq 0$, for each admissible and fixed value of L we will be able to regard $(L,M^*)_{\alpha a}$ as an extensor of range α : 0 to M^* whose full-range contraction with $V^{a(\alpha)}$ is the (M^*-M+L)th derivative of EVL for the given value of L. For fixed M^*, if $M^* \geq M$, there are M+1 of these extensors $(L,M^*)_{\alpha a}$, one for each value of L in the set 0, 1, . . . , M. For fixed M^*, if $M^* \leq M$, there are M^*+1 of these extensors $(L,M^*)_{\alpha a}$, one for each value of L in the set $M - M^*$, . . . ,M (since it is required that $L \geq M - M^*$ and $L \leq M$).

26. Construction of the one-parameter family $\mathcal{S}(E_{\alpha a})$. The following theorem will be used to provide a means for the construction from $E_{\alpha a}$ through the extensors $(L,M^*)_{\alpha a}$ additional, reduced-range extensors:

Theorem (26.1). If $E_{\alpha a}$ is an extensor defined over the range α : 0 to M, and if $L_{\alpha a} = \binom{\alpha}{L}^{-1} E_{(\alpha-L)a}$ for α : L to M + L, then $L_{\alpha a}$ is an extensor defined over the range α : L to M + L.

Conversely, if $E_{\alpha a}$ is an extensor defined over the range α : L to M + L, and if $T_{\alpha a} = \binom{\alpha+L}{L} E_{(\alpha+L)a}$ for α : 0 to M, then $T_{\alpha a}$ is an extensor defined over the range α : 0 to M.

Comment: Because the method of proof is straightforward, the proof is given in outline form.

Proof (Outline):

$$L_{\alpha a} = \binom{\alpha}{L}^{-1} \sum_{\rho=\alpha-L}^{M} E_{\rho r} X^{\rho r}_{(\alpha-L)a} \qquad (\text{for } \alpha : L \text{ to } M + L)$$

$$= \binom{\alpha}{L}^{-1} \sum_{\rho=\alpha-L}^{M} E_{\rho r} \binom{\rho}{\alpha-L} X_a^{r(\rho-\alpha+L)}$$

$$= \sum_{\rho=\alpha-L}^{M} \binom{\alpha}{L}^{-1} \binom{\rho}{\alpha-L} \binom{\rho+L}{\alpha}^{-1} E_{\rho r} X^{(\rho+L)r}_{\alpha a}$$

$$= \sum_{\rho=\alpha-L}^{M} \binom{\rho+L}{L}^{-1} E_{\rho r} X^{(\rho+L)r}_{\alpha a}$$

$$= \sum_{\sigma=\alpha}^{M+L} \binom{\sigma}{L}^{-1} E_{(\sigma-L)r} X^{\sigma r}_{\alpha a} = \sum_{\sigma=L}^{M+L} \binom{\sigma}{L}^{-1} E_{(\sigma-L)r} X^{\sigma r}_{\alpha a}$$

$$= \sum_{\sigma=L}^{M+L} L_{\sigma r} X^{\sigma r}_{\alpha a} \qquad \text{for } \alpha : L \text{ to } M + L.$$

Proof of converse (Outline):

$$T_{\alpha a} = \binom{\alpha+L}{L} \sum_{\rho=L}^{M+L} E_{\rho r} X^{\rho r}_{(\alpha+L)a}$$

$$= \binom{\alpha+L}{L} \sum_{\rho=L}^{M+L} E_{\rho r} \binom{\rho}{\alpha+L} X_a^{r(\rho-\alpha-L)}$$

$$= \sum_{\rho=L}^{M+L} \binom{\alpha+L}{L}\binom{\rho}{\alpha+L}\binom{\rho-L}{\alpha}^{-1} E_{\rho r} X_{\alpha a}^{(\rho-L)r}$$

$$= \sum_{\rho=L}^{M+L} \binom{\rho}{L} E_{\rho r} X_{\alpha a}^{(\rho-L)r} = \sum_{\sigma=0}^{M} \binom{\sigma+L}{L} E_{(\sigma+L)r} X_{\alpha a}^{\sigma r}$$

$$= \sum_{\sigma=0}^{M} T_{\sigma r} x_{\alpha a}^{\sigma r} \qquad \text{for } \alpha : 0 \text{ to } M.$$

Comment: It is seen that there is a countable infinity of extensors in the collection $\{(L,M^*)_{\alpha a} | 0 \leq L \leq M,\ L \geq M - M^*,\ M^* \geq 0\}$. Specifically, there is at least one extensor for each nonnegative value of M^*. We now note that Theorem (26.1) will show how to construct from this set of extensors $(L,M^*)_{\alpha a}$ other extensors $(L,M',\phi)_{\alpha a}$ of the arbitrary reduced range $\alpha : \phi$ to M', where $0 \leq \phi \leq M'$. The formula for $(L,M',\phi)_{\alpha a}$ will be expressed so as to describe the extensor $(L,M')_{\alpha a}$ when $\phi = 0$. A definition for the family $\mathcal{S}(E_{\alpha a})$ in the one-parameter case will then be adopted as indicated.

Let us now invoke Theorem (26.1) to insert additional extensors, specifically extensors of reduced range, into the family of extensors $(L,M^*)_{\alpha a}$. To the stipulations giving a range of values of L and M* such that $(L,M^*)_{\alpha a}$ is defined, namely $M > 0$, $M^* \geq 0$, $0 \leq L \leq M$, $M^* \geq M - L$, let us introduce the nonnegative integers M' and ϕ according to $M^* = M' - \phi$, where $0 \leq \phi \leq M'$. It follows, so long as $M' - \phi \geq M - L$, that $(L,M' - \phi)_{\alpha a}$ as defined by (25.1) is an extensor of range $\alpha : 0$ to $M' - \phi$. But if we adopt the definition

$$(26.1) \qquad (L,M',\phi)_{\alpha a} = \binom{\alpha}{\phi}^{-1} (L,M'-\phi)_{(\alpha-\phi)a} \quad ,$$

$\alpha : \phi$ to M', then by Theorem (26.1), $(L,M',\phi)_{\alpha a}$ is an extensor of range $\alpha : \phi$ to M'. Thus we have a formula for a set of extensors $(L,M',\phi)_{\alpha a}$ which encompasses the set of extensors $(L,M^*)_{\alpha a}$, i.e., when $\phi = 0$, we have $(L,M',0)_{\alpha a} = (L,M')_{\alpha a}$. We then have the following theorem:

Theorem (26.2). Given the absolute excovariant extensor $E_{\alpha a}$ of range $\alpha : 0$ to M with $M > 0$. Let L, M', and ϕ be arbitrary nonnegative integers such that $M' \geq M - L + \phi$, $0 \leq L \leq M$, $0 \leq \phi \leq M'$, and let

$D = M' - \phi - M + L$. Then $(L,M',\phi)_{\alpha a}$ are the components of an absolute excovariant extensor of the reduced range $\alpha : \phi$ to M', where

$$(L,M',\phi)_{\alpha a} = \begin{cases} \binom{\alpha}{\phi}^{-1} \sum_{\mu=D+\phi-\alpha}^{D} \binom{D}{\mu}\binom{\alpha+L-D-\phi+\mu}{L} E_{(\alpha+L-D-\phi+\mu)a}{}^{(\mu)} \\ \qquad \text{for } 0 \leq L \leq M,\ \alpha : \phi \text{ to } D + \phi\ ; \\ \binom{\alpha}{\phi}^{-1} \sum_{\mu=0}^{D} \binom{D}{\mu}\binom{\alpha+L-D-\phi+\mu}{L} E_{(\alpha+L-D-\phi+\mu)a}{}^{(\mu)} \\ \qquad \text{for } 0 \leq L < M,\ \alpha : D + \phi + 1 \text{ to } M'. \end{cases}$$

Comment: Any linear combination of extensors $(L,M',\phi)_{\alpha a}$ with identical range $\alpha : \phi$ to M' in Theorem (26.2) is an extensor. This suggests the introduction of the sets $\mathcal{M}(M',\phi)$ and $\mathcal{S}(E_{\alpha a})$ given by the following:

Definition (26.3). Assume there is given an extensor $E_{\alpha a}$ of range $\alpha : 0$ to M whose components are functions of class c^M. The set $\mathcal{M}(M',\phi)$ is the collection of all linear combinations of the extensors $(L,M',\phi)_{\alpha a}$ having the range $\alpha : \phi$ to M' with ϕ and M' fixed, $0 \leq \phi \leq M'$, and L variable but subject to the limitations

$L \geq M - M' + \phi$, $0 \leq L \leq M$. The family $\mathcal{S}(E_{\alpha a})$ is the union of all of the sets $\mathcal{M}(M',\phi)$, $M' \geq \phi \geq 0$.

Observation: There is a countable infinite number of extensors in $\mathcal{S}(E_{\alpha a})$, at least one extensor for each admissible pair ϕ and M'. Further, $E_{\alpha a}$ is in $\mathcal{S}(E_{\alpha a})$.

Remark: To illustrate the structure of the extensors $(L,M',\phi)_{\alpha a}$ let us assume $M = 3$, and let us choose $M' = 4$ and $\phi = 1$. Then L may assume the permissible fixed values 0, 1, 2, or 3. The associated extensors $(0,4,1)_{\alpha a}$, $(1,4,1)_{\alpha a}$, $(2,4,1)_{\alpha a}$, and $(3,4,1)_{\alpha a}$ are given rank-wise by equation (26.2).

$$
(26.2)\quad \begin{cases}
 & \alpha=1 & \alpha=2 & \alpha=3 & \alpha=4 \\
(0,4,1)_{\alpha a}: & E_{0a} & \frac{1}{2}E_{1a} & \frac{1}{3}E_{2a} & \frac{1}{4}E_{3a} \\
(1,4,1)_{\alpha a}: & E_{1a}{}' & \frac{1}{2}E_{1a}+E_{2a}{}' & \frac{2}{3}E_{2a}+E_{3a}{}' & \frac{3}{4}E_{3a} \\
(2,4,1)_{\alpha a}: & E_{2a}{}'' & E_{2a}{}'+\frac{3}{2}E_{3a}{}'' & \frac{1}{3}E_{2a}+2E_{3a}{}' & \frac{3}{4}E_{3a} \\
(3,4,1)_{\alpha a}: & E_{3a}{}''' & \frac{3}{2}E_{3a}{}'' & E_{3a}{}' & \frac{1}{4}E_{3a} .
\end{cases}
$$

We also note that, although it is true that if an extensor is in $\mathcal{S}(E_{\alpha a})$ then its ranks are linear combinations of parameter derivatives of the quantities E_{0a}, E_{1a}, . . . , E_{Ma}, the converse is not true. For example, if M = 2, the extensor $A_{\alpha a}$ of range α : 0 to 1, where $A_{0a} = E_{0a} - E_{2a}''$, $A_{1a} = E_{1a} - 2E_{2a}'$ is an extensor whose ranks are linear combinations of parameter derivatives of the quantities E_{0a}, E_{1a}, E_{2a}, and yet $A_{\alpha a}$ is not in $\mathcal{S}(E_{\alpha a})$. On the other hand, if the extensor $A_{\alpha a}$ is imbedded in the range α : 0 to 2 to form the extensor $B_{\alpha a}$ according to $B_{0a} = A_{0a}$, $B_{1a} = A_{1a}$, $B_{2a} = 0$, then $B_{\alpha a}$ is in $\mathcal{S}(E_{\alpha a})$ when M = 2.

Remark: It can be shown that a two-parameter extensor $((L_1,L_2),M^*;Q)_{(\alpha_1,\alpha_2)a}$ can be defined such that its contraction with $V^{a(\alpha_1,\alpha_2)}$ over the full range $\alpha_1 \geq 0$, $\alpha_2 \geq 0$, $\alpha_1 + \alpha_2 \leq M^*$ is a (D_1,D_2) derivative of a two-parameter analog to EVL, namely the invariant

$$\sum_{\alpha_1+\alpha_2\geq L_1+L_2}^{\alpha_1+\alpha_2\leq M} \binom{\alpha_1}{L_1}\binom{\alpha_2}{L_2} E_{(\alpha_1,\alpha_2)a} V^{a(\alpha_1-L_1,\alpha_2-L_2)} ,$$

along all class C^{M*} surfaces $x^a = x^a(u^1,u^2)$. The set of two-parameter generalized Kawaguchi extensors $((L_1,L_2),M^*;Q)_{(\alpha_1,\alpha_2)a}$ can then, with the aid of a two-parameter version of Theorem (26.1), be augmented to form the two-parameter family $\mathcal{S}(E_{\alpha a})$ of fundamental extensors associated with the two-parameter extensor $E_{(\alpha_1,\alpha_2)a}$ of range $\alpha_1 \geq 0$, $\alpha_2 \geq 0$, $\alpha_1 + \alpha_2 \leq M$.

CHAPTER V

CERTAIN LINEARITY PROPERTIES WITHIN $\mathcal{S}(E_{\alpha a})$

27. Subrealms of $\mathcal{S}(E_{\alpha a})$ and their dimensionality. In both the one-parameter and the two-parameter cases $\mathcal{S}(E_{\alpha a})$ has the closure property $E_{\alpha a} \in \mathcal{S}(E_{\alpha a})$. However, in neither case does it necessarily follow that $E_{\alpha a} \in \mathcal{S}(A_{\alpha a})$, given $A_{\alpha a} \in \mathcal{S}(E_{\alpha a})$. This observation raises questions concerning the linear structure of the family $\mathcal{S}(E_{\alpha a})$. Our purpose will not be to provide an exhaustive examination of the linearity properties of $\mathcal{S}(E_{\alpha a})$, but only to develop certain obvious properties sufficiently in order to apply systematically the family $\mathcal{S}(E_{\alpha a})$ to relevant problems in the calculus of variations.

In the one-parameter case an extensor $A_{\alpha a}$ is in $\mathcal{S}(E_{\alpha a})$ if and only if it is expressible as a linear combination of extensors $(L,M',\phi)_{\alpha a}$ of Theorem (26.2) having identical range $\alpha : \phi$ to M' such that $L \geq 0$, $M - M' + \phi \leq L \leq M$, $M' \geq 0$, $0 \leq \phi \leq M'$. Because of their prominence in the foundation of the one-parameter family $\mathcal{S}(E_{\alpha a})$, a first thought might be to regard the entire set of extensors $(L,M',\phi)_{\alpha a}$

of Theorem (26.2) as a basis for the one-parameter family $\mathcal{S}(E_{\alpha a})$.

However, since linear combinations are defined only for extensors of identical range, we will adopt what is perhaps a more appropriate procedure--the partitioning of $\mathcal{S}(E_{\alpha a})$ into subsets $\mathcal{M}$ called 'subrealms' of $\mathcal{S}(E_{\alpha a})$ (see Definition (8.8))--by fixing ϕ and M' in the one-parameter case. Having performed this partition, we will define a 'primary basis' for a subrealm $\mathcal{M}$ with fixed ϕ and M' in the one-parameter case as consisting of the set of extensors $(L,M',\phi)_{\alpha a}$ of Theorem (26.2), with the same fixed ϕ and M', obtained by taking all $L \geq 0$ such that $M - M' + \phi \leq L \leq M$.

Definition (27.1). The <u>subrealm $\mathcal{M}(M',\phi)$</u> is the subset of the family $\mathcal{S}(E_{\alpha a})$ of one-parameter extensors such that $A_{\alpha a} \epsilon \mathcal{M}(M',\phi)$ if and only if, for ϕ and M' given and fixed, $A_{\alpha a}$ is an extensor in $\mathcal{S}(E_{\alpha a})$ of the reduced range $\phi \leq \alpha \leq M'$.

Comment: Clearly, $\mathcal{S}(E_{\alpha a})$ can be regarded as the union of all its subrealms. Furthermore, each of these subrealms obeys the axioms of an algebraic linear Vector space by the usual definitions of the operations: (1) addition of extensors and (2) scalar multiplication, provided the concept 'a linearly independent set of Vectors in an algebraic linear Vector space' when applied to a set of extensors of identical range in $\mathcal{S}(E_{\alpha a})$ is interpreted as follows:

Definition (27.2). Given a family $\mathfrak{F}$ of admissible parameterized arcs with curve parameter t, with each arc passing through the fixed points P_1 and P_2 for $t = t_1$ and $t = t_2$, respectively. Given also a set $\mathcal{E}$ of extensors $(A_1)_{\alpha a}$, $(A_2)_{\alpha a}$, . . . , $(A_L)_{\alpha a}$ in $\mathcal{S}(E_{\alpha a})$ of identical range (say $\alpha : \emptyset$ to M') and defined at each point P(t) of every parameterized arc in $\mathfrak{F}$ satisfying $t_1 \leq t \leq t_2$. The set $\mathcal{E}$ is said to be a <u>linearly dependent set of extensors in $\mathcal{S}(E_{\alpha a})$ for the family $\mathfrak{F}$ of parameterized arcs and for the interval</u> $t_1 \leq t \leq t_2$ if and only if there exists a set of constants C_1, C_2, . . . , C_L such that the extensor $\sum_{K=0}^{L} C_K(A_K)_{\alpha a}$ vanishes in at least one (and hence in every) admissible coordinate system (x) at every point P(t) such that $t_1 \leq t \leq t_2$ of every parameterized arc in $\mathfrak{F}$. Otherwise, the set $\mathcal{E}$ is said to be a <u>linearly independent set of extensors in $\mathcal{S}(E_{\alpha a})$ for the family $\mathfrak{F}$ of parameterized arcs and for the interval</u> $t_1 \leq t \leq t_2$.

Remark: Even after linear independence is defined for a set of extensors defined over a family of parameterized arcs, the concept 'subrealm of the family $\mathcal{A}(E_{\alpha a})$' differs from that of 'manifold of an algebraic linear Vector space.' The distinction is that $\mathcal{A}(E_{\alpha a})$ does not have all of the closure properties of an algebraic linear Vector space, in that addition of extensors of unlike range remains undefined.

Definition (27.3). Let there be given (1) a family $\mathcal{F}$ of parameterized arcs and (2) an interval I: $t_1 \leq t \leq t_2$, with respect to which items the set $B(M',\phi)$ is linearly independent, where $B(M',\phi)$ contains the K extensors of the set $\{(L,M',\phi)_{\alpha a} \mid L \geq 0,\ M + \phi - M' \leq L \leq M\}$ of Theorem (26.2). Here, M' and ϕ are fixed, and

$$K \overset{\text{def}}{\equiv} \begin{cases} M' - \phi + 1 & \text{if } M > M' - \phi \\ M + 1 & \text{if } M \leq M' - \phi \end{cases} .$$

The extensors of $B(M',\phi)$ are each defined over the reduced range $\phi \leq \alpha \leq M'$. Then $B(M',\phi)$ will be called the <u>primary basis</u> for the subrealm $\mathcal{M}(M',\phi)$ in $\mathcal{A}(E_{\alpha a})$ for the family $\mathcal{F}$ and the interval I.

Comment: The statement that the subrealm $\mathcal{M}(M',\phi)$ contains just the set of all extensors which are linear combinations over its own primary basis is, of course, equivalent to the statement that $B(M', \emptyset)$ spans $\mathfrak{M}(M', \emptyset)$. But in the one-parameter case a domain of definition was specified involving a family $\mathfrak{F}$ of parameterized arcs and an interval $t_1 \leq t \leq t_2$ on an auxiliary t-axis with respect to which domain the set $B(M', \emptyset)$ was assumed to be a linearly independent set in the sense of Definition (27.2). This suggests that $B(M', \emptyset)$ can be regarded as a basis in the sense of linear algebra. Certain cases in which the set $B(M', \emptyset)$ is a linearly dependent set have been investigated separately by the author.

It is noteworthy that any alternate subset of extensors of $\mathfrak{M}(M', \emptyset)$ other than $B(M', \emptyset)$ having the property that it spans $\mathfrak{M}(M', \emptyset)$ and is a linearly independent set in the sense of Definition (27.2) may also be regarded as a basis for $\mathfrak{M}(M', \emptyset)$ in the sense of linear algebra. A basis for a subrealm $\mathfrak{M}$ of $\mathfrak{L}(E_{\alpha a})$ other than the primary basis will be called a secondary basis for $\mathfrak{M}$.

Now it is clear that an alternate basis for the subrealm $\mathfrak{M}(M', \emptyset)$ other than the primary basis $B(M', \emptyset)$ also contains $M' - \emptyset + 1$

extensors if $M > M' - \phi$, and M+1 extensors if $M \leq M' - \phi$. Accordingly, it is helpful to introduce the concept 'dimension of a subrealm' as in the following:

Definition (27.4). The <u>dimension of the subrealm $\mathcal{M}(M',\phi)$</u> of extensors of the one-parameter family $\mathcal{S}(E_{\alpha a})$ is the unique number K of extensors in a basis for $\mathcal{M}(M',\phi)$, where $K = M' - \phi + 1$ if $M > M' - \phi$ and $K = M + 1$ if $M \leq M' - \phi$.

We now proceed to impose physical dimensionality on components of extensors in $\mathcal{S}(E_{\alpha a})$ in order to show that the partitioning procedure corresponds to an alternate viewpoint that a linear combination may be formed only among extensors with rank-wise physical dimensional agreement. There is a countable infinity of distinct subrealms of $\mathcal{S}(E_{\alpha a})$ in both the one-parameter and the two-parameter cases. There is one and only one subrealm for each admissible pair M',ϕ in the one-parameter case. We see by the formula of Theorem (26.2) that, in the one-parameter case, there is a subrealm (given by $M' = M$, $\phi = 0$) in $\mathcal{S}(E_{\alpha a})$ of special interest in that it contains $E_{\alpha a}$ (for the case $L = 0$).

Let us be given the complete coordinate transformation $x^a = x^a(\bar{x})$; $\bar{x}^r = \bar{x}^r(x)$ connecting the domains D and $\bar{D}$ of auxiliary coordinate spaces one-to-one reversibly. We now first impose the requirement that each coordinate variable of x and $\bar{x}$ will have the physical dimensions of length. It follows that the quantities X^a_r are dimensionless. Next, we further require that the parameter t have dimensions of time. We note that, if in the x-coordinate system the components of the extensor $A_{\alpha a}$, $A_{\alpha a} \in \mathcal{S}(E_{\alpha a})$, have dimensions given by dim $A_{\alpha a} = L^c M^d T^{\alpha - k}$ where c, d, k are independent of α, then in a second $\bar{x}$-coordinate system the components $A_{\rho r}$ have dimensions given by dim $A_{\rho r} = L^c M^d T^{\rho - k}$, where L, M, T refer to units of length, mass, time. This previous statement follows by $A_{\rho r} = \sum_\alpha A_{\alpha a} X^{\alpha a}_{\rho r} = \sum_\alpha A_{\alpha a} \binom{\alpha}{\rho} X_r^{a(\alpha - \rho)}$, whence dim $A_{\rho r} = (\text{dim } A_{\alpha a}) \cdot T^{\rho - \alpha} = L^c M^d T^{\alpha - k} T^{\rho - \alpha} = L^c M^d T^{\rho - k}$.

Definition (27.5). Let $A_{\alpha a}$ be a one-parameter absolute excovariant extensor in the family $\mathcal{S}(E_{\alpha a})$ of fundamental extensors associated with the one-parameter absolute excovariant extensor $E_{\alpha a}$, of range α : 0 to M. Let $A_{\alpha a}$ have the range $\alpha : \phi, \phi+1, \ . \ . \ . \ , K$ with $K \geq \phi \geq 0$.

If dim $A_{\alpha a} = L^{c}M^{d}T^{\alpha-K}$ with c, d, k independent of α, then we refer to the triplet (c, d, α - k) as the <u>dimensional index d(α)</u> of the symbol $A_{\alpha a}$, and the numbers c, d, α - k as the <u>dimensional components</u> of d(α).

Note: This definition may be generalized for the case where $A_{\alpha a}$ is an extensor in the family $\mathcal{S}(E_{\alpha a})$ generated by the two-parameter extensor $E_{\alpha a}$.

Theorem (27.1). Let $A_{\alpha a}$ be a one-parameter absolute exco-variant extensor in $\mathcal{S}(E_{\alpha a})$ of range $\alpha : \emptyset, \emptyset+1, \ldots, K$ with $K \geq \emptyset \geq 0$. Assume that for each $\alpha : \emptyset, \emptyset+1, \ldots, K$ the dimensional indices of the different terms of $A_{\alpha a}$ agree. Then for each $\alpha : \emptyset, \emptyset+1, \ldots, K - 1$ the dimensional index d(α+1) of $A_{(\alpha+1)a}$ agrees in the first two dimensional components and exceeds by one in the third component the dimensional index d(α) of $A_{\alpha a}$.

Proof: By the extensor transformation formula it follows that for each $\alpha : \emptyset, \emptyset+1, \ldots, K$ we have

$$A_{\alpha a} = \sum_{\rho=\alpha}^{K} A_{\rho r} X^{\rho r}_{\alpha a} = \sum_{\rho=\alpha}^{K} \binom{\rho}{\alpha} A_{\rho r} X^{r(\rho-\alpha)}_{a} .$$

Suppose that the dimensional index of each term of $A_{\alpha a}$ (for an arbitrary α such that $\emptyset \leq \alpha \leq K$) is given by

$$d(\alpha) = (\ell, m, \alpha - k) .$$

Then $\dim A_{\alpha a} = (\dim A_{\rho r}) \cdot T^{\alpha-\rho}$, so that $\dim A_{\rho r} = T^{\rho-\alpha} \cdot \dim A_{\alpha a}$. Now, again by the extensor transformation formula,

$$A_{(\alpha+1)a} = \sum_{\rho=\alpha+1}^{K} A_{\rho r} X^{\rho r}_{(\alpha+1)a} = \sum_{\rho=\alpha+1}^{K} \binom{\rho}{\alpha+1} A_{\rho r} X^{r(\rho-\alpha-1)}_{a} ,$$

$\alpha : \emptyset, \emptyset+1, \ldots, K-1$, so that

$$\dim A_{(\alpha+1)a} = \dim A_{\rho r} T^{\alpha+1-\rho}$$

$$= \left[T^{\rho-\alpha} \dim A_{\alpha a} \right] \cdot T^{\alpha+1-\rho}$$

$$= T^{+1} \dim A_{\alpha a} .$$

Then from

$$\dim A_{\alpha a} = L^{\ell} M^{m} T^{\alpha-k}$$

we have

$$\dim A_{(\alpha+1)a} = \left[L^{\ell} M^{m} T^{\alpha-k} \right] \cdot T^{+1}$$

$$= L^{\ell} M^{m} T^{\alpha-k+1} .$$

Here, then, $d(\alpha) = (\ell, m, \alpha - k)$ implies that

$$d(\alpha+1) = (\ell, m, \alpha - k + 1)$$

and the proof is completed.

Note: The version of this theorem for two-parameter extensors may also be proved. We proceed to examine the subject of alternate bases for subrealms in $\mathcal{S}(E_{\alpha a})$ *via* the concept of Kawaguchi extensors.

Definition (27.6). Let $\mathcal{M}(M', \emptyset)$ be the subrealm of the one-parameter family $\mathcal{S}(E_{\alpha a})$ consisting of extensors of range $\alpha : \emptyset, \emptyset+1, \ldots, M'$. The extensor $K_{\alpha a} \in \mathcal{M}(M', \emptyset)$ is said to be a Kawaguchi-type extensor in $\mathcal{M}(M', \emptyset)$ if and only if (1) $K_{\alpha a} \not\equiv 0$ for some α such that $\emptyset \leq \alpha \leq M'$, and (2) there exists an R satisfying $\emptyset \leq R \leq M'$ such that $K_{\alpha a} = \binom{R}{\alpha} K_{Ra}{}^{\cdot(R-\alpha)}$ for each α satisfying $\emptyset \leq \alpha \leq M'$.

Remark: The convention of assigning the value zero to a symbol consisting of a negative differentiation applied to a base quantity assures us that $K_{\alpha a}$ is defined for each value of α satisfying $\emptyset \leq \alpha \leq M'$, an interval which may include α's for which $\alpha > R$. Actually, an assignment of any numerical value to $K_{Ra}{}^{\cdot(R-\alpha)}$ for $\alpha > R$ is satisfactory, for multiplication by $\binom{R}{\alpha}$, which vanishes for $\alpha > R$, causes $K_{\alpha a}$ to vanish for each $\alpha > R$.

Comment: It is seen at once that Definition (27.6) is not vacuous. In the one-parameter case every subrealm $\mathfrak{M}(M',\phi)$ of $\mathcal{S}(E_{\alpha a})$ contains a Kawaguchi-type extensor, namely $(M,M',\phi)_{\alpha a}$, in its primary basis. If $M > M' - \phi$, we will find $M' - \phi + 1$ Kawaguchi-type extensors $K_{\alpha a} = \binom{R}{\alpha} K_{Ra}{}^{\cdot(R-\alpha)}$, one for each value of R such that $\phi \leq R \leq M'$. If $M \leq M' - \phi$, we will find M+1 Kawaguchi-type extensors $K_{\alpha a} = \binom{R}{\alpha} K_{Ra}{}^{\cdot(R-\alpha)}$, one for each value of R such that $0 \leq R \leq M$.

These observations follow from the conjecture that the number of Kawaguchi-type extensors (excluding scalar multiples) in a subrealm is equal to the dimension of that subrealm. Alternately stated, it is our present assertion that a set of Kawaguchi-type extensors in $\mathfrak{M}$ of the type of Definition (27.6), having the property of linear independence in the sense of Definition (27.2), will constitute a secondary basis for a subrealm $\mathfrak{M}$. We proceed to discuss for the one-parameter case an iterative procedure called the (D,K)-process. The (D,K)-process uses the primary basis $B(M',\phi)$ to derive, for a subrealm $\mathcal{M}(M',\phi)$ of $\mathcal{S}(E_{\alpha a})$, a secondary basis $B_K(M',\phi)$ consisting of Kawaguchi-type extensors $(DK)_{\alpha a}[R,M',\phi]$, which are constructed from certain extensors $D_{\alpha a}[R,M',\phi]$ called difference extensors. The first argument R in the square brackets in the symbols $(DK)_{\alpha a}[R,M',\phi]$ and $D_{\alpha a}[R,M',\phi]$ indicates that R is the nonvanishing tensor rank number of these extensors in $\mathcal{M}(M',\phi)$. The (D,K)-process proceeds as follows:

First, take the primary basis extensor $(L^*,M',\phi)_{\alpha a}$ (alternately labeled $D_{\alpha a}[M',M',\phi]$), where L^* is the lowest admissible value which may be assigned L such that $(L,M',\phi)_{\alpha a}$ is a primary basis extensor in $\mathcal{M}(M',\phi)$. Thus $L^* = 0$ if $M + \phi - M' < 0$, $L^* = M + \phi - M'$ if $M + \phi - M' \geq 0$. Next, note that a scalar c may be chosen so as to produce agreement with $(L^*,M',\phi)_{\alpha a}$ in the nonvanishing tensor rank M' of the extensor $(DK)_{\alpha a}[M',M',\phi]$, defined by $(DK)_{\alpha a}[M',M',\phi] \overset{\text{def}}{\equiv} c(M,M',\phi)_{\alpha a}$, where $(M,M',\phi)_{\alpha a}$ is the Kawaguchi-type primary basis extensor in $\mathcal{M}(M',\phi)$. Here, $c = \binom{M}{M'-\phi}$ if $M - M' + \phi \geq 0$, $c = 1$ if $M - M' + \phi < 0$. With this actual choice for c, form $(DK)_{\alpha a}[M',M',\phi]$, then subtract

$(DK)_{\alpha a}[M',M',\emptyset]$ from $(L^*,M',\emptyset)_{\alpha a}$. Note that the resulting extensor $D_{\alpha a}[M'-1,M',\emptyset]$ has nonvanishing tensor rank $M' - 1$, assuming, of course, $M' \geq 1$. Next, construct from the difference extensor $D_{\alpha a}[M'-1,M',\emptyset]$ the Kawaguchi-type extensor $(DK)_{\alpha a}[M'-1,M',\emptyset]$ according to $(DK)_{\alpha a}[M'-1,M',\emptyset] \stackrel{\text{def}}{\equiv} \binom{M'-1}{\alpha} D_{(M'-1)a}[M'-1,M',\emptyset]^{(M'-1-\alpha)}$ for α : 0 to $M' - 1$; $(DK)_{M'a}[M'-1,M',\emptyset] \stackrel{\text{def}}{\equiv} 0$. Continuing, define $D_{\alpha a}[M'-2,M',\emptyset]$ (assuming $M' \geq 2$) by $D_{\alpha a}[M'-2,M',\emptyset] \stackrel{\text{def}}{\equiv} D_{\alpha a}[M'-1,M',\emptyset] - (DK)_{\alpha a}[M'-1,M',\emptyset]$ for α : 0 to $M' - 2$, $D_{(M'-1)a}[M'-2,M',\emptyset] \stackrel{\text{def}}{\equiv} 0$, $D_{M'a}[M'-2,M',\emptyset] \stackrel{\text{def}}{\equiv} 0$. Finally, after sufficient iterations, obtain a completely reduced Kawaguchi-type extensor $(DK)_{\alpha a}[L^{**},M',\emptyset]$, where $L^{**} = \emptyset$ if $M > M' - \emptyset$, $L^{**} = M' - M$ if $M \leq M' - \emptyset$. The set $\{(DK)_{\alpha a}[L,M',\emptyset] | L = L^{**}, L^{**}+1, \ldots, M'\}$ of Kawaguchi-type extensors is conjectured to be a basis for $\mathcal{M}(M',\emptyset)$. Further, it is easy to show that any such Kawaguchi-type extensor $(DK)_{\alpha a}[L,M',\emptyset]$, $L^{**} \leq L \leq M'$, has the property that for arbitrary class $C^{M'-\emptyset}$ tensors V^a the reduced-range extensor contraction $\sum_{\alpha=\emptyset}^{M'} \binom{\alpha}{\emptyset}(DK)_{\alpha a}[L,M',\emptyset]V^{a(\alpha-\emptyset)}$ is a perfect derivative of order $L - \emptyset$ with respect to the curve parameter t of the tensor contraction

$\binom{L}{\emptyset}V^a(DK)_{La}[L,M',\emptyset]$.

To illustrate with a first example, let us present a primary basis $B(4,1)$ and a secondary basis $B_K(4,1)$ for the subrealm $\mathcal{M}(4,1)$ in $\mathcal{S}(E_{\alpha a})$ when $M = 3$ (and $M' = 4, \emptyset = 1$). The primary basis $B(4,1)$ (see equation (26.2)) is given rank-wise by:

	$\alpha=1$	$\alpha=2$	$\alpha=3$	$\alpha=4$
$(0,4,1)_{\alpha a}$:	E_{0a}	$\frac{1}{2}E_{1a}$	$\frac{1}{3}E_{2a}$	$\frac{1}{4}E_{3a}$
$(1,4,1)_{\alpha a}$:	E_{1a}'	$\frac{1}{2}E_{1a}+E_{2a}'$	$\frac{2}{3}E_{2a}+E_{3a}'$	$\frac{3}{4}E_{3a}$
$(2,4,1)_{\alpha a}$:	E_{2a}''	$E_{2a}'+\frac{3}{2}E_{3a}''$	$\frac{1}{3}E_{2a}+2E_{3a}'$	$\frac{3}{4}E_{3a}$
$(3,4,1)_{\alpha a}$:	E_{3a}'''	$\frac{3}{2}E_{3a}''$	E_{3a}'	$\frac{1}{4}E_{3a}$

A secondary basis $B_K(4,1)$ for $\mathcal{M}(4,1)$ is given rank-wise by:

	$\alpha=1$	$\alpha=2$	$\alpha=3$	$\alpha=4$
$(DK)_{\alpha a}[4,4,1]$:	E_{3a}'''	$\frac{3}{2}E_{3a}''$	E_{3a}'	$\frac{1}{4}E_{3a}$
$(DK)_{\alpha a}[3,4,1]$:	$E_{2a}''-3E_{3a}'''$	$E_{2a}'-3E_{3a}''$	$\frac{1}{3}E_{2a}-E_{3a}'$	0
$(DK)_{\alpha a}[2,4,1]$:	$E_{1a}'-2E_{2a}''+3E_{3a}'''$	$\frac{1}{2}E_{1a}-E_{2a}'+\frac{3}{2}E_{3a}''$	0	0
$(DK)_{\alpha a}[1,4,1]$:	$E_{0a}-E_{1a}'+E_{2a}''-E_{3a}'''$	0	0	0

Here, $\sum_{\alpha=1}^{4}\binom{\alpha}{1}(DK)_{\alpha a}[L,4,1]V^{a(\alpha-1)}$ is $(E_{3a}V^a)'''$ for L = 4; $\left[(E_{2a} - 3E_{3a}')V^a\right]''$ for L = 3; $\left[(E_{1a} - 2E_{2a}' + 3E_{3a}'')V^a\right]'$ for L = 2; and simply $(E_{0a} - E_{1a}' + E_{2a}'' - E_{3a}''')V^a$ for L = 1. The Kawaguchi-type bases for subrealms in $\mathcal{S}(E_{\alpha a})$ lend themselves perhaps more readily than the primary bases for application to the calculus of variations. Their importance in this regard suggests the inclusion of further examples.

We present a primary basis B(2,1) and a secondary basis $B_K(2,1)$ for the subrealm $\mathcal{M}(2,1)$ in $\mathcal{S}(E_{\alpha a})$ when M = 3 (and M$'$= 2,$\emptyset$ = 1). The primary basis B(2,1) is given rank-wise by:

	$\alpha=1$	$\alpha=2$
$(2,2,1)_{\alpha a}$:	E_{2a}	$\frac{3}{2}E_{3a}$
$(3,2,1)_{\alpha a}$:	E_{3a}'	$\frac{1}{2}E_{3a}$

A secondary basis $B_K(2,1)$ of Kawaguchi-type extensors for $\mathcal{M}(2,1)$ is given rank-wise by:

	$\alpha=1$	$\alpha=2$
$(DK)_{\alpha a}[2,2,1]$:	$3E_{3a}'$	$\frac{3}{2}E_{3a}$
$(DK)_{\alpha a}[1,2,1]$:	$E_{2a}-3E_{3a}'$	0

Here, $\sum_{\alpha=1}^{2}\binom{\alpha}{1}(DK)_{\alpha a}[L,2,1]V^{a(\alpha-1)}$ is $(3E_{3a}V^a)'$ for $L = 2$; and simply $(E_{2a}-3E_{3a}')V^a$ for $L = 1$. We conclude Section 27 with a final illustration of alternate bases of subrealms of the one-parameter family $\mathcal{S}(E_{\alpha a})$.

The primary basis $B(3,0)$ for the subrealm $\mathcal{M}(3,0)$ in $\mathcal{S}(E_{\alpha a})$ when $M = 2$ (and $M'=3, \phi=0$) is given rank-wise by:

	$\alpha=0$	$\alpha=1$	$\alpha=2$	$\alpha=3$
$(0,3,0)_{\alpha a}$:	E_{0a}'	$E_{0a}+E_{1a}'$	$E_{1a}+E_{2a}'$	E_{2a}
$(1,3,0)_{\alpha a}$:	E_{1a}''	$2E_{1a}'+2E_{2a}''$	$E_{1a}+4E_{2a}'$	$2E_{2a}$
$(2,3,0)_{\alpha a}$:	E_{2a}'''	$3E_{2a}''$	$3E_{2a}'$	E_{2a}

A secondary basis $B_K(3,0)$ for the same subrealm $\mathcal{M}(3,0)$ in the one-parameter family $\mathcal{S}(E_{\alpha a})$ when $M = 2$ is the set of Kawaguchi-type extensors given rank-wise by:

	$\alpha=0$	$\alpha=1$	$\alpha=2$	$\alpha=3$
$(DK)_{\alpha a}[3,3,0]$:	E_{2a}'''	$3E_{2a}''$	$3E_{2a}'$	E_{2a}
$(DK)_{\alpha a}[2,3,0]$:	$E_{1a}''-2E_{2a}'''$	$2E_{1a}'-4E_{2a}''$	$E_{1a}-2E_{2a}'$	0
$(DK)_{\alpha a}[1,3,0]$:	$E_{0a}'-E_{1a}''+E_{2a}'''$	$E_{0a}-E_{1a}'+E_{2a}''$	0	0

Here, $\sum_{\alpha=0}^{3} (DK)_{\alpha a}[L,3,0]V^{a(\alpha)}$ is $(E_{2a}V^a)'''$ for L = 3; $\left[(E_{1a}-2E_{2a}')V^a\right]''$ for L = 2; and $\left[(E_{0a}-E_{1a}'+E_{2a}'')V^a\right]'$ for L = 1.

28. <u>The subrealm $\mathcal{M}_0$ containing $E_{\alpha a}$</u>. Of particular interest in the one-parameter case will be the subrealm $\mathcal{M}_0 (=\mathcal{M}(M,0))$ given by $\phi = 0$, $M' = M$. The primary basis B_0 $(=B(M,0))$ for this subrealm is given by

$$B_0 = \{(L,M,0)_{\alpha a} | L = 0, 1, \ldots, M\} ,$$

where

$$(28.1) \qquad (L,M,0)_{\alpha a} = \sum_{\mu=0}^{L} \binom{L}{\mu}\binom{\alpha+\mu}{L} E_{(\alpha+\mu)a}^{(\mu)} ,$$

$\alpha = 0, 1, \ldots, M$ for each $L = 0, 1, \ldots, M$. Note that $E_{\alpha a}$ is contained as the element $(0,M,0)_{\alpha a}$ in the primary basis for $\mathcal{M}_0$.

It can also be shown that for arbitrary class c^M tensors V^a, if $L \neq 0$, the extensor contraction $\sum_{\alpha=0}^{M} (L,M,0)_{\alpha a} V^{a(\alpha)}$ is a perfect derivative of order L with respect to the curve parameter t, given by

$$(28.2) \qquad \sum_{\alpha=0}^{M} (L,M,0)_{\alpha a} V^{a(\alpha)} = \left[\sum_{\delta=L}^{M} \binom{\delta}{L} E_{\delta a} V^{a(\delta-L)}\right]^{(L)} .$$

It is of further interest to note the alternate basis $B_{K,0} = \{(DK)_{\alpha a}[L,M,0] \mid L = 0, 1, \ldots, M\}$, where we have $D_{\alpha a}[M,M,0] \overset{\text{def}}{\equiv} E_{\alpha a}$, $(DK)_{\alpha a}[M,M,0] \overset{\text{def}}{\equiv} \binom{M}{\alpha} E_{Ma}{}^{(M-\alpha)}$ for α : 0 to M, $D_{\alpha a}[M-L,M,0] \overset{\text{def}}{\equiv} D_{\alpha a}[M-L+1,M,0] - (DK)_{\alpha a}[M-L+1,M,0]$ for $L = 0, 1, \ldots, M-1$, and $(DK)_{\alpha a}[M-L,M,0] \overset{\text{def}}{\equiv} \binom{M-L}{\alpha} D_{(M-L)a}[M-L,M,0]^{(M-L-\alpha)}$ for $L = 0, 1, \ldots, M-1$. Here, each $(DK)_{\alpha a}[M-L,M,0]$ with $L = 0, 1, \ldots, M-1$ has the property that $\sum_{\alpha=0}^{M} (DK)_{\alpha a}[M-L,M,0] V^{a(\alpha)}$ is a perfect derivative of order $M - L$.

CHAPTER VI

THE SET OF SYNGE TENSORS ASSOCIATED WITH $E_{\alpha a}$ AND A CLASS OF VARIATIONAL INTEGRALS

29. Kawaguchi base tensors from $\mathcal{S}(E_{\alpha a})$. In pursuit of our objective of constructing covariant tensors from a generative absolute excovariant extensor $E_{\alpha a}$, we seek to obtain these tensors in certain ways from the extensors in $\mathcal{S}(E_{\alpha a})$. This suggests at once the extraction of a set S_K of Kawaguchi base tensors associated with $E_{\alpha a}$ from the set of extensors in $\mathcal{S}(E_{\alpha a})$ having a pure Kawaguchi structure. Following this extraction a set S_H of Hamiltonian-type tensors associated with $\mathcal{S}(E_{\alpha a})$ will be derived, and shown to be the same set as the set S_K of Kawaguchi base tensors associated with $E_{\alpha a}$. Here, application of the idea that the vanishing of a Hamiltonian tensor describes a normalizing arc for an extensor-generalized calculus of variations integral provides a simple generalized definition in which a Hamiltonian-type tensor is associated with each of the (absolute excovariant) extensors in $\mathcal{S}(E_{\alpha a})$, including those of reduced range. This set $S_H(=S_K)$ of

tensors associated with $E_{\alpha a}$ will then be shown identical to the set S_S of Synge tensors associated with $E_{\alpha a}$, derived in Section 13. Its Hamiltonian interpretation will immediately provide a desired new variational significance for each member of the set S_S of Synge tensors. In Section 30 a family of variational integrals of the type $\int_{t_1}^{t_2} \sum_{\alpha=\emptyset}^{M'} \binom{\alpha}{\emptyset} A_{\alpha a} V^{a(\alpha-\emptyset)} dt$, $A_{\alpha a} \in \mathcal{S}(E_{\alpha a})$, is given. Normalizing arcs for certain of these integrals are found to be describable not only by the vanishing of a Hamiltonian-type tensor obtained from $\mathcal{S}(E_{\alpha a})$, but also by (a) the vanishing of a pure Kawaguchi-type extensor in $\mathcal{S}(E_{\alpha a})$, or by (b) equality of the integrand extensor $A_{\alpha a}$ with a particular linear combination of generalized Kawaguchi extensors $(L,M',\emptyset)_{\alpha a}$ of Chapter IV. At present we consider a pair of definitions involving Kawaguchi base tensors derived from $\mathcal{S}(E_{\alpha a})$:

Definition (29.1). Let $\mathcal{M}$ be a subrealm of $\mathcal{S}(E_{\alpha a})$, consisting of extensors of range over the set $\mathcal{R} = \{\alpha | \emptyset \le \alpha \le M'\}$ in the one-parameter case. Let $K_{\alpha a}$ be a Kawaguchi-type extensor in $\mathcal{M}$ (so that we know by Definition (27.6) that $K_{\alpha a} \not\equiv 0$ for some $\alpha \in \mathcal{R}$). Then if, in the one-parameter case R, $R \in \mathcal{R}$, is the nonvanishing tensor rank of $K_{\alpha a}$ so that $K_{\alpha a}$ has the structure of Definition (27.6), namely

$$K_{\alpha a} = \binom{R}{\alpha} K_{Ra}{}^{(R-\alpha)}$$

for each $\alpha \in \mathcal{R}$, the quantities K_{Ra} will be said to be the Kawaguchi base tensor associated with $K_{\alpha a}$. The number R will be called the formative rank index for $K_{\alpha a}$.

Comment: We have already seen that every subrealm of $\mathcal{S}(E_{\alpha a})$ contains at least one Kawaguchi-type extensor. Thus, in the comment following Definition (27.6) we noted that the subrealm $\mathcal{M}(M',\emptyset)$ contains the Kawaguchi-type extensor $(M,M',\emptyset)_{\alpha a}$ in its primary basis.

Definition (29.2). Let $\mathcal{M}$ be a subrealm of $\mathcal{S}(E_{\alpha a})$. A tensor T_a is said to be a $K\mathcal{M}$ base tensor if and only if there exists a Kawaguchi-type extensor $K_{\alpha a} \in \mathcal{M}$ with Kawaguchi base tensor T_a. A tensor T_a is said to be a $K\mathcal{S}$ base tensor if and only if there exists a subrealm $\mathcal{M}$ of $\mathcal{S}(E_{\alpha a})$ such that T_a is a $K\mathcal{M}$ base tensor.

Definition (29.3). In the one-parameter case, each of the subrealms $\mathcal{M}(M'+1,\emptyset)$ and $\mathcal{M}(M',\emptyset+1)$ will be said to be successive to the subrealm $\mathcal{M}(M',\phi)$.

Remark: For convenience in generating the set S_K of $K\mathcal{A}$ base tensors, we will use a recursion formula for primary basis elements in successive subrealms in $\mathcal{A}(E_{\alpha a})$. Theorem (29.1) is such a recursion formula.

Theorem (29.1). Let B_P be the primary basis for the subrealm $\mathcal{M}(P,0)$ of extensors of range α : 0 to P, $P \geq 0$, and let B_{P+1} be the primary basis for the subrealm $\mathcal{M}(P+1,0)$ of extensors of range α : 0 to P + 1. Then, for each L such that $(L,P+1,0)_{\alpha a} \in B_{P+1}$ and such that $(L,P,0)_{\alpha a} \in B_P$ (i.e., $M - P \leq L \leq M$ and $L \geq 0$) and for each α such that $0 \leq \alpha \leq P + 1$ we have

$$(L,P+1,0)_{\alpha a} = (L,P,0)_{(\alpha-1)a} + (L,P,0)_{\alpha a}'$$

(where $(L,P,0)_{(-1)a} \overset{\text{def}}{\equiv} 0$, and where $(L,P,0)_{\alpha a}' = 0$ for $\alpha = P + 1$, as $E_{\gamma a} \overset{\text{def}}{\equiv} 0$ for $\gamma > M$).

Proof: By Theorem (26.2), using $\binom{M'-M+L-\phi}{\mu} \overset{\text{def}}{\equiv} 0$ for $\mu < 0$ and $E_{(\alpha-M'+M+\mu)a}{}^{(\mu)} \overset{\text{def}}{\equiv} 0$ for $\mu < 0$, we have for $M' \geq 0$, $M' \geq M - L + \phi$,

$0 \leq L \leq M$, $0 \leq \phi \leq M'$ the result (for $\alpha : \phi$ to M')

$$(a)\ (L,M',\phi)_{\alpha a} = \binom{\alpha}{\phi}^{-1} \sum_{\mu=M'-M+L-\alpha}^{M'-M+L-\phi} \binom{M'-M+L-\phi}{\mu} \binom{\alpha-M'+M+\mu}{L} E_{(\alpha-M'+M+\mu)a}{}^{(\mu)} .$$

Accordingly, for $\alpha = 0$, the formula to be proved reduces to the identity $E_{La}{}^{(P+1-M+L)} = E_{La}{}^{(P+1-M+L)}$. For $\alpha = P + 1$ the formula to be proved becomes the identity $\binom{M}{L} E_{Ma} = \binom{M}{L} E_{Ma}$. For $1 \leq \alpha \leq P$ we have by (a) the result

$$(L,P,0)_{(\alpha-1)a} = \sum_{\mu=P-M+L-\alpha+1}^{P-M+L} \binom{P-M+L}{\mu} \binom{\alpha-1-P+M+\mu}{L} E_{(\alpha-1-P+M+\mu)a}{}^{(\mu)} ,$$

which becomes, by replacing the index μ by $\mu + 1$, the formula

$$(b)\ (L,P,0)_{(\alpha-1)a} = \sum_{\mu=P-M+L-\alpha}^{P-M+L} \binom{P-M+L}{\mu+1} \binom{\alpha-P+M+\mu}{L} E_{(\alpha-P+M+\mu)a}{}^{(\mu+1)} .$$

Here, we have utilized the fact that $\binom{P-M+L}{\mu+1} \equiv 0$ for $\mu > P - M + L - 1$ to use $P - M + L$ in place of $P - M + L - 1$ as the upper limit on the μ-summation.

Also by (a) we have, for $1 \leq \alpha \leq P$,

$$(c)\ (L,P,0)_{\alpha a}{}' = \sum_{\mu=P-M+L-\alpha}^{P-M+L} \binom{P-M+L}{\mu} \binom{\alpha-P+M+\mu}{L} E_{(\alpha-P+M+\mu)a}{}^{(\mu+1)} .$$

Therefore, by (b) and (c), and by $\binom{P-M+L}{\mu+1} + \binom{P-M+L}{\mu} = \binom{P-M+L+1}{\mu+1}$, we have,

for $1 \leq \alpha \leq P$,

$$(L,P,0)_{(\alpha-1)a} + (L,P,0)_{\alpha a}{}' = \sum_{\mu=P-M+L-\alpha}^{P-M+L} \binom{P-M+L+1}{\mu+1}\binom{\alpha-P+M+\mu}{L} E^{(\mu+1)}_{(\alpha-P+M+\mu)a},$$

which becomes, by replacing the index μ by $\mu - 1$, the result

$$\text{(d)} \quad (L,P,0)_{(\alpha-1)a} + (L,P,0)_{\alpha a}{}' = \sum_{\mu=P+1-M+L-\alpha}^{P+1-M+L} \binom{P+1-M+L}{\mu}\binom{\alpha-P-1+M+\mu}{L} E^{(\mu)}_{(\alpha-P-1+M+\mu)a}$$

for $1 \leq \alpha \leq P$, i.e.,

$$\text{(e)} \quad (L,P,0)_{(\alpha-1)a} + (L,P,0)_{\alpha a}{}' = (L,P+1,0)_{\alpha a}$$

for $1 \leq \alpha \leq P$. But we have proved (e) separately for the cases $\alpha = 0$ and $\alpha = P+1$, and the theorem follows.

Note: As a further convenience in obtaining Kawaguchi base tensors associated with $E_{\alpha a}$ we will relate the set of Kawaguchi-type extensors in a given subrealm $\mathfrak{M}$ of $\mathcal{S}(E_{\alpha a})$ to the set of primary basis extensors in $\mathfrak{M}$ by means of an explicit formula in Theorem (29.4). First, however, we will consider Theorems (29.2) and (29.3). Theorem (29.2) will show that, in a given subrealm $\mathfrak{M}$, there are no Kawaguchi-type extensors other than nonzero scalar multiples of extensors of the set $\{(DK)_{\alpha a}[R,M',\emptyset] \mid L^{**} \leq R \leq M'\}$ of Section 27. Then Theorem (29.3) will show that certain linear combinations of the primary basis extensors for $\mathcal{M}$ have a pure Kawaguchi structure. In Theorems (29.2), (29.3), and (29.4), L^* is defined, as usual, according to $L^* = 0$ if $M+\emptyset-M' < 0$, $L^* = M + \emptyset - M'$ if $M + \emptyset - M' \geq 0$. Also, L^{**} is defined, as usual, according to $L^{**} = M' - M$ if $M + \emptyset - M' < 0$, $L^{**} = \emptyset$ if $M + \emptyset - M' \geq 0$.

Theorem (29.2). An extensor in the subrealm $\mathcal{M}(M',\emptyset)$ of the one-parameter family $\mathcal{S}(E_{\alpha a})$ is a Kawaguchi-type extensor if and only if it is a nonzero scalar multiple of one of the extensors of the set $\{(DK)_{\alpha a}[R,M',\emptyset] \mid L^{**} \leq R \leq M'\}$.

Proof: We first note that the set $\{(DK)_{\alpha a}[R,M',\emptyset] \mid L^{**} \leq R \leq M'\}$ contains the same number of extensors as is contained in the primary basis $\{(L,M',\emptyset)_{\alpha a} \mid L^* \leq L \leq M\}$ for $\mathcal{M}(M',\emptyset)$. Because the set $\{(DK)_{\alpha a}[R,M',\emptyset] \mid L^{**} \leq R \leq M'\}$ contains exactly $M - L^* + 1$ extensors (as, by definition of L^{**} and L^*, we have $L^{**} = M' - M + L^*$) and is a linearly independent set in the sense of Definition (27.2), it follows that the set $\{(DK)_{\alpha a}[R,M',\emptyset]|R = L^{**}, L^{**}+1, . . . , M'\}$ is a basis for $\mathcal{M}(M',\emptyset)$. Consequently, if the extensor $A_{\alpha a}$ is in $\mathcal{M}(M',\emptyset)$, then there exists a set $\{C_R|R = L^{**},L^{**}+1, . . . , M'\}$ of scalars such that

$$A_{\alpha a} = \sum_{R=L^{**}}^{M'} C_R \, (DK)_{\alpha a}[R,M',\emptyset] \quad .$$

Now if $A_{\alpha a}$ is a Kawaguchi-type extensor, it must have a formative rank $\hat{R}$ such that

$$A_{\alpha a} = \begin{cases} \binom{\hat{R}}{\alpha} A_{\hat{R}a}^{(\hat{R}-\alpha)} & \text{for } \emptyset \le \alpha \le \hat{R}\ , \quad \hat{R} \le M'\ ; \\ 0 & \text{for } \hat{R}+1 \le \alpha \le M'\ ,\ \hat{R} < M'\ . \end{cases}$$

If $\hat{R} = M'$, then because $(DK)_{M'a}[R,M',\emptyset] = 0$ whenever $L^{**} \le R \le M' - 1$ (when $M' \ge 1$) we have

$$A_{M'a} = C_{M'}(DK)_{M'a}\,[M',M',\emptyset]\ ,$$

so that in order for $A_{\alpha a}$ to have Kawaguchi structure we must have

$$A_{\alpha a} = C_{M'}(DK)_{\alpha a}[M',M',\emptyset]\ , \qquad C_{M'} \ne 0\ ,$$

and the theorem follows for the case $\hat{R} = M'$.

If $\hat{R} = M' - 1$ (with $M' \geq 1$) and $A_{\alpha a}$ is a Kawaguchi-type extensor, then

$$A_{\alpha a} = \begin{cases} \binom{M'-1}{\alpha} A_{(M'-1)a}{}^{(M'-1-\alpha)} & \text{for } \emptyset \leq \alpha \leq M' - 1 \\ 0 & \text{for } \alpha = M' . \end{cases}$$

But $A_{M'a} = C_{M'}(DK)_{M'a}[M',M',\emptyset]$ and $(DK)_{M'a}[M',M',\emptyset] \not\equiv 0$. Accordingly, if $\hat{R} = M' - 1$ (so that $A_{M'a} = 0$), we must have $C_{M'} = 0$. We then have

$$A_{\alpha a} = \sum_{R=L^{**}}^{M'-1} C_R (DK)_{\alpha a}[R,M',\emptyset]$$

and

$$A_{(M'-1)a} = \sum_{R=L^{**}}^{M'-1} C_R (DK)_{(M'-1)a}[R,M',\emptyset] .$$

But $(DK)_{(M'-1)a}[R,M',\emptyset] = 0$ whenever $L^{**} \leq R \leq M' - 2$, $M' \geq 2$. Then, if $\hat{R} = M' - 1$, we have for the tensor rank $M' - 1$ of $A_{\alpha a}$ the formula

$$A_{(M'-1)a} = C_{M'-1}(DK)_{(M'-1)a}[M'-1,M',\emptyset] ,$$

so that in order for $A_{\alpha a}$ to have Kawaguchi structure we must have

$$A_{\alpha a} = C_{M'-1}(DK)_{\alpha a}[M'-1,M',\emptyset],\ C_{M'-1} \not\equiv 0,$$

and the theorem follows for the case $\hat{R} = M' - 1$.

In general, if $\hat{R} = M' - S$ for some S such that $0 \leq S \leq M' - L^{**}$, then we have, if $A_{\alpha a}$ is a Kawaguchi-type extensor,

$$A_{\alpha a} \equiv \begin{cases} \binom{M'-S}{\alpha} A_{(M'-S)a}^{(M'-S-\alpha)} & \text{for } \emptyset \leq \alpha \leq M' - S \text{ ;} \\ 0 & \text{for } M' - S + 1 \leq \alpha \leq M' (S>0). \end{cases}$$

But $A_{M'a} = C_{M'}(DK)_{M'a}[M',M',\emptyset]$ and $(DK)_{M'a}[M',M',\emptyset] \not\equiv 0$. Hence, because $A_{M'a} = 0$, we must have $C_{M'} = 0$. Then, we have $A_{(M'-1)a} = C_{M'-1}(DK)_{M'-1a}[M'-1,M',\emptyset]$ and $(DK)_{M'-1a}[M',M',\emptyset] \not\equiv 0$. Hence, because $A_{M'-1a} = 0$, we must have $C_{M'-1} = 0$. We may continue this argument to obtain $C_{M'} = C_{M'-1} = \ldots = C_{M'-S+1} = 0$ (when $S > 0$). We then have

$$A_{\alpha a} = \sum_{R=L^{**}}^{M'-S} C_R (DK)_{\alpha a}[R,M',\emptyset]$$

and

$$A_{(M'-S)a} = \sum_{R=L^{**}}^{M'-S} C_R (DK)_{(M'-S)a}[R,M',\emptyset] \quad .$$

But $(DK)_{(M'-S)a}[R,M',\emptyset] = 0$ whenever $L^{**} \leq R < M' - S$. Then, if $\hat{R} = M' - S$, we have for the tensor rank $M' - S$ of $A_{\alpha a}$ the formula

$$A_{(M'-S)a} = C_{M'-S}(DK)_{(M'-S)a}[M'-S,M',\emptyset] \quad ,$$

so that in order for $A_{\alpha a}$ to have Kawaguchi structure we must have

$$A_{\alpha a} = C_{M'-S}(DK)_{\alpha a}[M'-S, M', \emptyset],\ C_{M'-S} \neq 0 \quad ,$$

for any S such that $0 \leq S \leq M' - L^{**}$, and the theorem follows.

Theorem (29.3). Let $\mathfrak{M}$ be the subrealm $\mathfrak{M}(M', \emptyset)$ of the one-parameter family $\mathcal{S}(E_{\alpha a})$ consisting of extensors of range $\alpha : \emptyset$ to M'. Let $\{(L, M', \emptyset)_{\alpha a} \mid L \geq 0,\ M - M' + \emptyset \leq L \leq M\}$ be the primary basis for $\mathfrak{M}$. Let R satisfy $L^{**} \leq R \leq M'$, and let J be defined according to $J \overset{\text{def}}{\equiv} R - M' + M$ (so that $L^{**} - M' + M \leq J \leq M$). Then the extensor $R_{\alpha a}$ in $\mathfrak{M}$ that is given by

$$(29.1) \qquad R_{\alpha a} = \sum_{L=L^*}^{M} (-1)^{L-J} \binom{L}{J} (L, M', \emptyset)_{\alpha a}$$

for $\alpha : \emptyset$ to M' has the Kawaguchi structure

$$(29.2) \qquad R_{\alpha a} = \begin{cases} \binom{R}{\alpha} \left[\binom{R}{\emptyset}^{-1} S_a^{[J]} \right]^{(R-\alpha)} & \text{for } \alpha : \emptyset \text{ to } R,\ R \leq M' \ ; \\ 0 & \text{for } \alpha : R+1 \text{ to } M',\ R < M', \end{cases}$$

where $S_a^{[J]}$ is that Synge tensor of Section 13, equation (13.1), indicated by the bracketed superscript.

Remark: As noted, $L^{**} - M' + M \leq J \leq M$. But $L^{**} - M' + M = L^*$, and $L^* \geq 0$. Accordingly, $0 \leq J \leq M$. This last inequality guarantees that the symbol $S_a^{[J]}$ is well defined.

Proof of Theorem (29.3): Define the extensor $(KR)_{\alpha a}$ as identical to the right member of (29.2). Accordingly, the reduced range extensor contraction $\sum_{\alpha=\emptyset}^{M'} \binom{\alpha}{\emptyset}(KR)_{\alpha a} V^{a(\alpha-\emptyset)}$ can be written as the expression

$$\sum_{\alpha=\emptyset}^{R} \binom{\alpha}{\emptyset}\binom{R}{\alpha}\left[\binom{R}{\emptyset}^{-1} S_a^{[J]}\right]^{(R-\alpha)} V^{a(\alpha-\emptyset)} \quad .$$

Now by (13.1) we have

$$(29.3) \qquad S_a^{[J]} = \sum_{\beta=0}^{M'-R} (-1)^{\beta}\binom{\beta+J}{J} E_{(\beta+J)a}^{(\beta)} \quad .$$

Therefore, there follows

$$(29.4) \qquad \sum_{\alpha=\emptyset}^{M'} \binom{\alpha}{\emptyset}(KR)_{\alpha a} V^{a(\alpha-\emptyset)} = \sum_{\alpha=\emptyset}^{R} \binom{\alpha}{\emptyset}\binom{R}{\alpha}\binom{R}{\emptyset}^{-1} \sum_{\beta=0}^{M'-R} (-1)^{\beta} \binom{\beta+J}{J} E_{(\beta+J)a}^{(\beta+R-\alpha)} V^{a(\alpha-\emptyset)} \quad .$$

On the other hand, the extensor $R_{\alpha a}$ given by the right member of (29.1) has a reduced range contraction $\sum_{\alpha=\emptyset}^{M'} \binom{\alpha}{\emptyset} R_{\alpha a} V^{a(\alpha-\emptyset)}$ which can be written as

$$\sum_{\alpha=\emptyset}^{M'} \binom{\alpha}{\emptyset} \sum_{L=L^*}^{M} (-1)^{L-J}\binom{L}{J}(L,M',\emptyset)_{\alpha a} V^{a(\alpha-\emptyset)} \quad .$$

But by (26.1) we have

$$(29.5)\quad \sum_{\alpha=\phi}^{M'} \binom{\alpha}{\phi}(L,M',\phi)_{\alpha a} V^{a(\alpha-\phi)} = \sum_{\alpha=\phi}^{M'} (L,M'-\phi)_{(\alpha-\phi)a} V^{a(\alpha-\phi)} \quad .$$

By replacing the index α of summation in the right member of (29.5) by $\alpha + \phi$, we have that the right member of (29.5) becomes

$$\sum_{\alpha=0}^{M'-\phi} (L,M'-\phi)_{\alpha a} V^{a(\alpha)} \quad .$$

Now by Theorem (25.1) this last expression becomes

$$(EVL)^{(D)} \quad ,$$

where EVL is given by Definition (25.1) as

$$\sum_{\alpha=L}^{M} \binom{\alpha}{L} E_{\alpha a} V^{a(\alpha-L)} \quad ,$$

and where $D = M' - \phi - M + L$. Therefore, we have that $\sum_{\alpha=\phi}^{M'} \binom{\alpha}{\phi} R_{\alpha a} V^{a(\alpha-\phi)}$ can be written as

$$\sum_{L=L^*}^{M} (-1)^{L-J} \binom{L}{J} \sum_{\alpha=L}^{M} \binom{\alpha}{L} \left[E_{\alpha a} V^{a(\alpha-L)}\right]^{(D)} \quad .$$

Now by an expansion of $\left[E_{\alpha a} V^{a(\alpha-L)}\right]^{(D)}$ by the Leibnitz formula we obtain

$$(29.6)\quad \sum_{\alpha=\phi}^{M'} \binom{\alpha}{\phi} R_{\alpha a} V^{a(\alpha-\phi)} = \sum_{L=L^*}^{M} (-1)^{L-J} \binom{L}{J} \sum_{\alpha=L}^{M} \binom{\alpha}{L} \sum_{\gamma=0}^{D} \binom{D}{\gamma} E_{\alpha a}^{(D-\gamma)} V^{a(\alpha-L+\gamma)} \quad .$$

In order to show the identity of the left members of (29.4) and (29.6) it suffices to show the identity of respective coefficients of $V^{a(\epsilon)}$ in the right members of (29.4) and (29.6) for each ϵ such that $0 \leq \epsilon \leq M' - \phi$. Now the coefficient of $V^{\alpha(\epsilon)} (0 \leq \epsilon \leq M' - \phi)$ in the right member of (29.4) is

$$\binom{\epsilon+\phi}{\phi} \binom{R}{\epsilon+\phi} \binom{R}{\phi}^{-1} \sum_{\beta=0}^{M'-R} (-1)^{\beta} \binom{\beta+J}{J} E_{(\beta+J)a}^{(\beta+R-\epsilon-\phi)} \quad ,$$

which vanishes whenever $R - \phi < \epsilon \leq M' - \phi$, as $\binom{R}{\epsilon+\phi} \equiv 0$ for $\epsilon > R - \phi$. The coefficient of $V^{a(\epsilon)} (0 \leq \epsilon \leq M' - \phi)$ in the right member of (29.6) becomes, after a modification of the lower limit of the L-summation,

$$\sum_{L=J}^{M} (-1)^{L-J} \binom{L}{J} \sum_{\alpha=L}^{M} \binom{\alpha}{L} \binom{D}{\epsilon-\alpha+L} E_{\alpha a}^{(D-L-\epsilon+\alpha)} \quad .$$

Here, the lower limit L^* on the L-summation was replaced by J, as $\binom{L}{J} \equiv 0$ for $L < J$ and as $L^* \leq J$ followed from $L^{**} \leq R$, $L^{**} = L^* + M' - M$.

But now (for each fixed ϵ such that $0 \leq \epsilon \leq M' - \phi$) the identity of these last two expressions can be proved by showing, for these two expressions, the identity of respective coefficients of $E_{\delta a}^{(\delta+M'-M-\epsilon-\phi)}$ for each δ such that $J \leq \delta \leq M$. The coefficient of $E_{\delta a}^{(\delta+M'-M-\epsilon-\phi)}(J \leq \delta \leq M)$ in the expression for the coefficient of $V^{a(\epsilon)}(0 \leq \epsilon \leq M' - \phi)$ in the right member of (29.4) is

$$\binom{\epsilon+\phi}{\phi}\binom{R}{\epsilon+\phi}\binom{R}{\phi}^{-1}(-1)^{\delta-J}\binom{\delta}{J} \quad .$$

The coefficient of $E_{\delta a}^{(\delta+M'-M-\epsilon-\phi)}(J \leq \delta \leq M)$ in the expression for the coefficient of $V^{a(\epsilon)}(0 \leq \epsilon \leq M' - \phi)$ in the right member of (29.6) is

$$\sum_{L=J}^{M} (-1)^{L-J}\binom{L}{J}\binom{\delta}{L}\binom{D}{\epsilon-\delta+L} \quad .$$

By the substitution $\binom{L}{J}\binom{\delta}{L} = \binom{\delta}{J}\binom{\delta-J}{L-J}$ the proposed identity becomes

$$\text{(29.7)} \qquad \sum_{L=J}^{M} (-1)^{L-J}\binom{\delta}{J}\binom{\delta-J}{L-J}\binom{D}{\epsilon-\delta+L}$$

$$= \binom{\epsilon+\phi}{\phi}\binom{R}{\epsilon+\phi}\binom{R}{\phi}^{-1}\binom{\delta}{J}(-1)^{\delta-J} \quad .$$

Now by the substitution $\binom{\epsilon+\phi}{\phi}\binom{R}{\epsilon+\phi}\binom{R}{\phi}^{-1} = \binom{R-\phi}{R-\phi-\epsilon}$ and by the division of both members of (29.7) by $\binom{\delta}{J}$ the proposed identity becomes

$$(29.8) \qquad \sum_{L=J}^{M} (-1)^{L-J} \binom{\delta-J}{L-J}\binom{D}{\epsilon-\delta+L} = \binom{R-\phi}{R-\phi-\epsilon}(-1)^{\delta-J}.$$

Next, we replace the index L of summation in the left member of (29.8) by k according to $k = L - J$. After a modification of the upper limit of the k-summation, (29.8) becomes

$$(29.9) \qquad \sum_{k=0}^{\delta-J} (-1)^{k}\binom{\delta-J}{k}\binom{R+k-\phi}{M'-\phi-M-\epsilon+\delta} = \binom{R-\phi}{R-\phi-\epsilon}(-1)^{\delta-J} .$$

Here, the upper limit $M' - R$ on the k-summation was replaced by $\delta - J$, as $\binom{\delta-J}{k} \equiv 0$ for $k > \delta - J$, and as $M' - R \geq \delta - J$ followed from $\delta \leq M$. Now by the substitutions $n = \delta - J$, $x = R - \phi$, $r = M' - \phi - M - \epsilon + \delta$ the proposed identity (29.9) becomes

$$(29.10) \qquad \sum_{k=0}^{n} (-1)^{k}\binom{n}{k}\binom{x+k}{r} = (-1)^{n}\binom{x}{r-n} \qquad ,$$

which is a previously established identity involving binomial coefficients (see [26], page 24, eq. (3.47)). We may now conclude that the left members of (29.4) and (29.6) are identical, so that we have

$$(29.11) \qquad \sum_{\alpha=\phi}^{M'} \binom{\alpha}{\phi}(KR)_{\alpha a}V^{a(\alpha-\phi)} = \sum_{\alpha=\phi}^{M'} \binom{\alpha}{\phi}R_{\alpha a}V^{a(\alpha-\phi)}$$

for arbitrary class $C^{M'-\emptyset}$ tensors V^a. But for each α we may select a fixed pair β,b such that $V^a = \dfrac{\delta^\alpha_\beta \delta^a_b \binom{\alpha}{\emptyset}^{-1}(t-t_0)^{\alpha-\emptyset}}{(\alpha-\emptyset)!}$. By varying β,b over all ordered pairs such that $\emptyset \leq \beta \leq M'$, $1 \leq b \leq N$, we obtain the extensor equation $(KR)_{\beta b} = R_{\beta b}$, and the theorem is proved.

Comment: We will utilize Theorems (29.2) and (29.3) in the proof of Theorem (29.4). We have

Theorem (29.4). Let $\mathfrak{M}$ be the subrealm $\mathfrak{M}(M',\emptyset)$ of the one-parameter family $\mathcal{S}(E_{\alpha a})$ consisting of extensors of range $\alpha : \emptyset$ to M'. Let $\{(DK)_{\alpha a}[R,M',\emptyset] \mid L^{**} \leq R \leq M'\}$ be the set of Kawaguchi-type extensors in $\mathfrak{M}$ constructed by the (D,K)-process. For each R such that $L^{**} \leq R \leq M'$ there exists a scalar k_R, independent of α, such that the extensors of the set of Kawaguchi-type extensors in $\mathfrak{M}$ satisfy the following formula involving the primary basis extensors $(L,M',\emptyset)_{\alpha a}$ in $\mathfrak{M}$:

$$(29.12)\quad (DK)_{\alpha a}[R,M',\emptyset] = k_R \sum_{L=L^*}^{M} (-1)^{L-J}\binom{L}{J}(L,M',\emptyset)_{\alpha a}.$$

Remark: It can be proved that $k_R \equiv \binom{J}{L^*}$, but this result will not be needed to obtain the set S_K of Kawaguchi base tensors.

Proof of Theorem (29.4): We have by Theorem (29.3) that the right member of (29.12) has the Kawaguchi structure

$$\begin{cases} k_R\binom{R}{\alpha}\left[\binom{R}{\emptyset}^{-1} S_a^{[J]}\right]^{(R-\alpha)} & \text{for } \alpha : \emptyset \text{ to } R,\ R \leq M' ; \\ 0 & \text{for } \alpha : R+1 \text{ to } M',\ R < M' , \end{cases}$$

$J = R - M' + M$. Thus the right member of (29.12) is a Kawaguchi-type extensor in $\mathfrak{M}(M',\emptyset)$ with formative rank R.

Next, we have by Theorem (29.2) that the right member of (29.12), for each permissible fixed value of R, must be a nonzero scalar multiple of that member of the set $\{(DK)_{\alpha a}[S,M',\emptyset] \mid L^{**} \leq S \leq M'\}$ with formative rank R, namely the extensor $(DK)_{\alpha a}[R,M',\emptyset]$. Thus, for each R such that $L^{**} \leq R \leq M'$, there exists a nonzero scalar, say k_R, such that

$$(29.13)\quad (DK)_{\alpha a}\left[R,M',\emptyset\right] \equiv \begin{cases} k_R\binom{R}{\alpha}\left[\binom{R}{\emptyset}^{-1} S_a^{[J]}\right]^{(R-\alpha)} & \text{for } \alpha : \emptyset \text{ to } R, R \leq M'; \\ 0 & \text{for } \alpha : R+1 \text{ to } M', R < M'. \end{cases}$$

But, by Theorem (29.3), the right member of (29.13) is identical to the right member of (29.12), and the theorem is proved.

Note: We have seen that (i) if $M' > M + \phi$, there is a unique (save for scalar multiples) nonvanishing Kawaguchi-type extensor for each of the formative ranks $M' - M$ to M', but none with formative rank ϕ, $\phi+1$, . . . , $M' - M - 1$; (ii) if $M' \leq M + \phi$, there is a unique (save for scalar multiples) nonvanishing Kawaguchi-type extensor for each of the formative ranks ϕ to M'.

Remark: Recalling that there is no more than one (except for scalar multiples of the original) Kawaguchi-type extensor in a given subrealm $\mathfrak{M}$ with a given formative rank R, we will adopt the phraseology

T_a is the $K\mathfrak{M}$ base tensor for the formative rank R if there is a Kawaguchi-type extensor in $\mathfrak{M}$ based on T_a with formative rank R. The following theorem shows that the same tensor may be a $K\mathfrak{M}$ base tensor for many different subrealms $\mathfrak{M}$ in $\mathcal{S}(E_{\alpha a})$.

Theorem (29.5). If T_a is the $K\mathfrak{M}_1$ base tensor for the formative rank R (where $\mathfrak{M}_1$ is the subrealm $\mathfrak{M}(M',\phi)$ of extensors of range $\alpha : \phi$ to M'), then (i) T_a is also the $K\mathfrak{M}_2$ base tensor for the formative rank R (where $\mathfrak{M}_2$ is the subrealm $\mathfrak{M}(M',\phi+1)$ of extensors of range

$\alpha : \emptyset + 1$ to M'), and (ii) T_a is also the $K\mathcal{M}_3$ base tensor for the formative rank R (where $\mathcal{M}_3$ is the subrealm $\mathcal{M}(M'+1,\emptyset)$ of extensors of range $\alpha : \emptyset$ to $M' + 1$).

Observation: Consider the permissible formative ranks associated with $\mathcal{M}(M',\emptyset)$. If $M' \geq M + \emptyset$ in the one-parameter case, these ranks are $M' - M$, $M' - M + 1$, . . . , M'. If a tensor is the $K\mathcal{M}_0$ base tensor (where $\mathcal{M}_0 = \mathcal{M}(M',\emptyset)$) with formative rank R, it is also the $K\mathcal{M}_1$ base tensor (where $\mathcal{M}_1 = \mathcal{M}(M'-1,\emptyset)$ and $M' \geq 1$) with formative rank $R - 1$, where $R \geq 1$, and by induction it is the $K\mathcal{M}_S$ base tensor (where $\mathcal{M}_S = \mathcal{M}(M'-S,\emptyset)$ and $M' \geq S$) with formative rank $R - S$, where $R \geq S$, for each S such that $0 \leq S \leq M' - M$.

Furthermore, any tensor which is the $K\mathcal{M}$ base tensor for the subrealm $\mathcal{M}(M,\emptyset)$, $0 \leq \emptyset \leq M'$, with the formative rank R is also the $K\mathcal{M}$ base tensor for the subrealm $\mathcal{M}(M,0)$ with the same formative rank R. Now if $M' \leq M + \emptyset$ in the one-parameter case the permissible formative ranks are $\emptyset$, $\emptyset+1$, . . . , M', and if a tensor is the $K\mathcal{M}$ base tensor for the subrealm $\mathcal{M}(M',\emptyset)$ with formative rank R, it is the $K\mathcal{M}$ base tensor for the subrealm $\mathfrak{M}(M'+1,\emptyset)$ with formative rank $R + 1$, and by induction it is the $K\mathfrak{M}$ base tensor for the subrealm $\mathfrak{M}(M'+S,\emptyset)$ with formative rank $R + S$, for each S such that $0 \leq S \leq M - M'$, where $R + S \geq \emptyset$. The same tensor is the $K\mathfrak{M}$ base tensor for the subrealm $\mathfrak{M}(M,0)$ with formative rank $R + M - M'$, where $R + M - M' \geq 0$ follows from $R \geq \emptyset$, $M' \leq M + \emptyset$.

Therefore, if a tensor is a $K\mathfrak{M}$ base tensor for any subrealm $\mathfrak{M}$ of $\mathcal{S}(E_{\alpha a})$, it is a $K\mathfrak{M}$ base tensor for the particular subrealm $\mathfrak{M} = \mathfrak{M}(M,0)$ containing $E_{\alpha a}$. We conclude that the set S_K of Kawaguchi base tensors associated with $E_{\alpha a}$, i.e., the set S_K of $K\mathcal{S}$ base tensors in the one-parameter case is the set $\{(DK)_{Ra}[R,M,0] \mid 0 \leq R \leq M\}$, where $(DK)_{\alpha a}[R,M,0]$ is given by

$$(29.14) \qquad (DK)_{\alpha a}[R,M,0] = \sum_{L=0}^{M} (-1)^{L-R}\binom{L}{R}(L,M,0)_{\alpha a},$$

where the primary basis extensors $(L,M,0)_{\alpha a}$ are given by Theorem (26.2). Thus the set S_K of Kawaguchi base tensors associated with $E_{\alpha a}$ is given by

$$(DK)_{Ra}[R,M,0] = \sum_{L=0}^{M} (-1)^{L-R}\binom{L}{R} \sum_{\mu=0}^{L} \binom{L}{\mu}\binom{R+\mu}{L} E_{(R+\mu)a}{}^{(\mu)} \quad ,$$

$R = 0, 1, \ldots, M$, which may be rewritten as

$$(29.15) \qquad (DK)_{Ra}[R,M,0] = \sum_{\mu=0}^{M-R} (-1)^{\mu}\binom{R+\mu}{R} E_{(R+\mu)a}{}^{(\mu)} ,$$

$R = 0, 1, \ldots, M$.

This equivalence may be seen by verifying that

$$\sum_{L=0}^{M} (-1)^{L-R}\binom{L}{R}\binom{L}{\mu}\binom{R+\mu}{L} = (-1)^{\mu}\binom{\mu+R}{R}$$

is an identity for $0 \leq \mu \leq M - R$, $0 \leq R \leq M$, $M > 0$. But, by $\binom{L}{\mu}\binom{R+\mu}{L} = \binom{R+\mu}{R}\binom{R}{L-\mu}$, this reduces to the proposition

$$\sum_{L=0}^{M} (-1)^{L-R}\binom{L}{R}\binom{R}{L-\mu} = (-1)^{\mu} \quad ,$$

i.e.,

$$\sum_{L=R}^{R+\mu} (-1)^{L}\binom{L}{R}\binom{R}{L-\mu} = (-1)^{R+\mu} \quad ,$$

which is an identity involving binomial coefficients (see [20], p. 286, eq. (7.9.20)).

30. <u>A certain family of variational integrals and a set of Hamiltonian-type tensors from $\mathcal{L}(E_{\alpha a})$.</u> The extensor generalizations of

certain simple calculus of variations problems, given in [15] and [16], may be utilized to generate families of calculus of variations problems and associated variational integrals. Thus we assume:

(1) There is given an extensor $A_{\alpha a}$ in a subrealm $\mathfrak{M}(M',\phi)$ of extensors in $\mathcal{S}(E_{\alpha a})$ of range ϕ to M', with $M' > \phi$. (2) The components of $A_{\alpha a}$ are functions of $x, x', \ldots, x^{(P)}$, $P \geq M$, and are of class $c^{M'}$ for x in a connected region R' and for all values of the x-primes which satisfy $\Sigma(x'^{a})^2 > 0$. (3) There is given the interval $t_1 \leq t \leq t_2$ on an auxiliary t-axis, a pair of points P,Q in R' with coordinates denoted by x_1^a, x_2^a, and the set of arcs $x^a = x^a(t)$, $t_1 \leq t \leq t_2$, of class c^H, which lie entirely in R' and satisfy the boundary conditions $x^{(\alpha)a}(t_1) = x_1^{\alpha a}$, $x^{(\alpha)a}(t_2) = x_2^{\alpha a}$ (see [15]). Here, $0 \leq \alpha \leq M' - \phi - 1$ and H is the functional order of the expression

$$\sum_{\alpha=0}^{M-L^*} (-1)^{\alpha} \binom{\alpha+L^*}{L^*} E_{(\alpha+L^*)a}^{(\alpha)} ,$$

where L^* is given, as usual, by $L^* = \max\{0, M-M'+\phi\}$.

A simple extension of the extensor-generalization is the problem to find an admissible arc $C_0 : x^a = x_0^a(t)$ such that if

$C_v : x^a = x^a(t,v)$ is a set of admissible arcs with $x^a(t,0) = x^a_0(t)$,

then for $A_{\alpha a}$ evaluated along C_0,

$$(30.1) \qquad \int_{t_1}^{t_2} \sum_{\alpha=\phi}^{M'} \binom{\alpha}{\phi} A_{\alpha a} V^{a(\alpha-\phi)} dt = 0,$$

where the tensor V^a is a set of variation functions associated with the set of arcs C_v such that $V^{a(\Omega)}(t_1) = V^{a(\Omega)}(t_2) = 0$ for $\Omega = 0, 1, \ldots, M' - \phi - 1$.

Remark: Before proceeding to examine the class of such variational problems as $A_{\alpha a}$ sweeps out $\mathcal{A}(E_{\alpha a})$, let us observe an equivalency which occurs when $A_{\alpha a}$ is a primary basis extensor $(L,M',\phi)_{\alpha a}$. We have that

$$\int_{t_1}^{t_2} \sum_{\alpha=\phi}^{M'} \binom{\alpha}{\phi} (L,M',\phi)_{\alpha a} V^{a(\alpha-\phi)} dt$$

can be replaced by

$$\int_{t_1}^{t_2} \sum_{\alpha=\phi}^{M'} \binom{\alpha}{\phi} \left[\binom{\alpha}{\phi}^{-1} (L,M'-\phi,0)_{(\alpha-\phi)a}\right] V^{a(\alpha-\phi)} dt,$$

or the equivalent

$$\int_{t_1}^{t_2} \sum_{\beta=0}^{M'-\phi} (L,M'-\phi,0)_{\beta a} V^{a(\beta)} dt \quad .$$

Here, the quantities $(L,M'-\emptyset,0)_{\beta a}$ are the components of an extensor of range β : 0 to $M' - \emptyset$. Accordingly, the extended extensor-generalized calculus of variations problem based on the reduced range contraction $\sum_{\alpha=\emptyset}^{M'} \binom{\alpha}{\emptyset}(L,M',\emptyset)_{\alpha a}V^{a(\alpha-\emptyset)}$ as integrand can be replaced by a standard extensor-generalized calculus of variations problem with a full range extensor contraction, namely $\sum_{\beta=0}^{M'-\emptyset} (L,M'-\emptyset,0)_{\beta a}V^{a(\beta)}$, as integrand. Normalizing arcs (arcs along which (30.1) is satisfied) for the reduced range type problem are described by the vanishing of the Hamiltonian tensor for an associated full range type problem. Therefore, the following definition is suggested:

Definition (30.1). The Hamiltonian-type tensor associated with the reduced range extensor $(L,M',\emptyset)_{\alpha a}$ of Theorem (26.2) in the subrealm $\mathfrak{M}(M',\emptyset)$ of extensors of range α : $\emptyset$ to M', is the Hamiltonian tensor (see [15]) associated with the full range extensor $(L,M'-\emptyset,0)_{\alpha a}$ in the subrealm $\mathfrak{M}(M'-\emptyset,0)$ of extensors of range α : 0 to $M' - \emptyset$.

Comment: Similarly, we define the Hamiltonian-type tensor associated with a linear combination of the $(L,M',\emptyset)_{\alpha a}$ as the same linear combination of the associated Hamiltonian-type tensors. The following theorem shows that most of the extensors $(L,M',\emptyset)_{\alpha a}$ have a vanishing Hamiltonian-type tensor:

Theorem (30.1). The Hamiltonian-type tensor associated with $(L,M',\emptyset)_{\alpha a}$, given by the expression

$$\sum_{\alpha=0}^{M'-\emptyset} (-1)^{\alpha}(L,M'-\emptyset,0)_{\alpha a}^{(\alpha)} ,$$

vanishes unless we have both $M - M' + \emptyset \geq 0$ and $L = M - M' + \emptyset$.

Proof: By Theorem (25.1) we have

$$\sum_{\alpha=0}^{M'-\emptyset} (L,M'-\emptyset,0)_{\alpha a} V^{a(\alpha)} = (EVL)^{(M'-\emptyset-M+L)} ,$$

where

$$EVL = \sum_{\alpha=L}^{M} \binom{\alpha}{L} E_{\alpha a} V^{a(\alpha-L)} .$$

Accordingly, we have for the case $M' - \emptyset - M + L > 0$ the result

$$(30.2) \quad \int_{t_1}^{t_2} \sum_{\alpha=0}^{M'-\emptyset} (L,M'-\emptyset,0)_{\alpha a} V^{a(\alpha)} dt = \left[\sum_{\alpha=L}^{M} \binom{\alpha}{L} E_{\alpha a} V^{a(\alpha-L)} \right]^{(M'-\emptyset-M+L-1)} \Bigg|_{t_1}^{t_2} .$$

Now the right member of (30.2) consists of a set of terms, each of which contains one of $V^a \Big|_{t_1}^{t_2}$, $V^{a'} \Big|_{t_1}^{t_2}$, . . . , $V^{a(M'-\emptyset-1)} \Big|_{t_1}^{t_2}$ as a factor.

But in the extensor-generalized problem based upon the extensor $(L,M',\phi)_{\alpha a}$ we have that the variation tensor V^a satisfies $V^{a(\Omega)}(t_1) = V^{a(\Omega)}(t_2) = 0$ for $\Omega = 0, 1, \ldots, M' - \phi - 1$. Accordingly, the left member of (30.2) vanishes, so that

$$(30.3) \qquad \int_{t_1}^{t_2} \sum_{\alpha=\phi}^{M'} \binom{\alpha}{\phi} (L,M',\phi)_{\alpha a} V^{a(\alpha-\phi)} dt \equiv 0 \quad .$$

Therefore, whenever $M' - \phi - M + L > 0$, i.e., whenever $L > M - M' + \phi$, we have that (30.3) is satisfied along all arcs. Then, because all arcs are normalizing arcs for $(L,M',\phi)_{\alpha a}$, it follows that the Hamiltonian-type tensor associated with $(L,M',\phi)_{\alpha a}$ vanishes identically unless it is not the case that $L > M - M' + \phi$. But by definition of $(L,M',\phi)_{\alpha a}$ we require both $L \geq 0$ and $L \geq M - M' + \phi$. Hence, the Hamiltonian-type tensor associated with $(L,M',\phi)_{\alpha a}$ vanishes identically unless we have both $M - M' + \phi \geq 0$ and $L = M - M' + \phi$.

Remark: In case $M - M' + \phi \geq 0$ and $L = M - M' + \phi$ we may write an explicit form for the Hamiltonian-type tensor $H_a^{[M-M'+\phi]}$ associated with $(L,M,\phi)_{\alpha a}$. Thus for $L = M - M' + \phi$, with $M - M' + \phi \geq 0$, the expression for this Hamiltonian-type tensor, namely

$$\sum_{\alpha=0}^{M'-\phi} (-1)^{\alpha}(L,M'-\phi,0)_{\alpha a}{}^{(\alpha)} \quad ,$$

becomes

$$\sum_{\alpha=0}^{M'-\phi} (-1)^{\alpha}(M-M'+\phi,M'-\phi,0)_{\alpha a}{}^{(\alpha)}.$$

But for the case at hand $L^* = M - M' + \phi$. For this case, by Theorem (26.2), we have

$$(L^*,M'-\phi,0)_{\alpha a} = \binom{\alpha+L^*}{L^*} E_{(\alpha+L^*)a} \quad ,$$

so that the Hamiltonian-type tensor $H_a{}^{[L^*]}$ associated with $(L,M',\phi)_{\alpha a}$ when $M - M' + \phi \geq 0$ and $L = M - M' + \phi$ is given by

$$(30.4) \qquad H_a{}^{[L^*]} = \sum_{\alpha=0}^{M-L^*} (-1)^{\alpha}\binom{\alpha+L^*}{L^*} E_{(\alpha+L^*)a}{}^{(\alpha)} \quad .$$

Now we obtain an expression for the Hamiltonian-type tensor associated with a general extensor $A_{\alpha a}$ such that $A_{\alpha a} \in \mathcal{S}(E_{\alpha a})$.

First, we observe that, if $A_{\alpha a} \in \mathcal{M}(M',\phi)$, then $A_{\alpha a}$ is expressible as a linear combination of the primary basis extensors for $\mathcal{M}(M',\phi)$ according to

$$A_{\alpha a} = \sum_{L=L^*}^{M} K_L (L,M',\phi)_{\alpha a}$$

with the constants K_L appropriately chosen. Here, $\mathcal{M}(M',\phi)$ is an arbitrary subrealm of $\mathcal{S}(E_{\alpha a})$, and L^* is given once more by $L^* = M - M' + \phi$ if $M - M' + \phi \geq 0$, $L^* = 0$ if $M - M' + \phi < 0$. Next, we note that the left member of (30.1) becomes

$$\int_{t_1}^{t_2} \sum_{\alpha=\phi}^{M'} \binom{\alpha}{\phi} \sum_{L=L^*}^{M} K_L (L,M',\phi)_{\alpha a} V^{a(\alpha-\phi)} dt \quad ,$$

or the equivalent

$$\sum_{L=L^*}^{M} K_L \left[\int_{t_1}^{t_2} \sum_{\alpha=\phi}^{M'} \binom{\alpha}{\phi} (L,M',\phi)_{\alpha a} V^{a(\alpha-\phi)} dt \right] .$$

Then by Theorem (30.1), equation (30.3), we conclude that this last expression vanishes identically if $M - M' + \phi < 0$ (so that $L^* = 0$), but reduces to

$$K_{M-M'+\phi} \left[\int_{t_1}^{t_2} \sum_{\alpha=\phi}^{M'} \binom{\alpha}{\phi} (M-M'+\phi,M',\phi)_{\alpha a} V^{a(\alpha-\phi)} dt \right]$$

if $M - M' + \phi \geq 0$ (so that $L^* = M - M' + \phi$). By (30.4) the bracket in this last expression has a nonvanishing Hamiltonian-type tensor $H_a^{[M-M'+\phi]}$. Thus we have

Theorem (30.2). Let $A_{\alpha a} \epsilon \mathfrak{M}(M', \phi)$. Let $\{K_L | L^* \leq L \leq M\}$ be chosen so that $A_{\alpha a} = \sum_{L=L^*}^{M} K_L(L, M', \phi)_{\alpha a}$. The Hamiltonian-type tensor associated with $A_{\alpha a}$ vanishes identically unless we have both $M - M' + \phi \geq 0$ and $K_{M-M'+\phi} \neq 0$, in which case the Hamiltonian-type tensor associated with $A_{\alpha a}$ is $K_{M-M'+\phi} H_a^{[M-M'+\phi]}$ (see eq. (30.4)).

Remark: It follows that, if $M - M' + \phi < 0$, all admissible arcs C are normalizing arcs for $A_{\alpha a}$. If $M - M' + \phi \geq 0$, either (a) all admissible arcs C are normalizing arcs for $A_{\alpha a}$, or (b) only arcs C_0 for which $H_a^{[M-M'+\phi]}$ vanishes are normalizing arcs for $A_{\alpha a}$, according as $K_{M-M'+\phi}$ (a) vanishes, or (b) does not vanish.

Now let us determine the set S_H of tensors such that A_a is in S_H if and only if there exists an extensor $A_{\alpha a}$ in $\mathcal{S}(E_{\alpha a})$ with associated nonvanishing Hamiltonian-type tensor A_a. We introduce first Definitions (30.2) and (30.3).

Definition (30.2). A tensor H_a is said to be an <u>$H\mathfrak{M}$ tensor</u> if, given the subrealm $\mathfrak{M}$ of $\mathcal{S}(E_{\alpha a})$, there exists an extensor in $\mathfrak{M}$ whose associated Hamiltonian-type tensor is H_a.

Definition (30.3). A tensor H_a is said to be an <u>$H\mathcal{S}$tensor</u> if there exists a subrealm $\mathcal{M}$ of $\mathcal{S}(E_{\alpha a})$ such that H_a is an $H^{\mathcal{M}}$tensor.

Comment: By Theorem (30.2) the set S_H of tensors, i.e., the set of all $H\mathcal{S}$tensors, is given by successively assigning to $M' - \phi$ in (30.4) the values 0, 1, . . . , M. Alternately, as by the substitution $\gamma = M - M' + \phi$ equation (30.4) becomes

$$(30.5) \qquad H_a^{[\gamma]} = \sum_{\alpha=0}^{M-\gamma} (-1)^{\alpha} \binom{\alpha+\gamma}{\gamma} E_{(\alpha+\gamma)a}^{(\alpha)} \qquad ,$$

we have that S_H is given by successively assigning to γ in (30.5) the values 0, 1, . . . , M.

It is seen by comparison with (29.15) and (13.1) that the set S_H of tensors satisfies $S_H = S_K = S_S$, where S_K is the set of Kawaguchi base tensors associated with $E_{\alpha a}$, and where S_S is the set of Synge tensors associated with $E_{\alpha a}$. (The analogous statement is valid for the two-parameter case.) We will sometimes refer to $S_H (= S_K = S_S)$ henceforth simply as the set S of tensors associated with $E_{\alpha a}$. We note that S contains the Hamiltonian-type tensor $H_a^{[0]}$ associated with $E_{\alpha a}$ itself.

Therefore, the quest for normalizing arcs over the class of variational integrals

$$\int_{t_1}^{t_2} \sum_{\alpha=\phi}^{M'} \binom{\alpha}{\phi} A_{\alpha a} V^{a(\alpha-\phi)} dt$$

as M' and ϕ run over the values $M' = 0, 1, \ldots ; \phi = 0, 1, \ldots, M'$, and as all $A_{\alpha a}$ such that $A_{\alpha a} \in \mathcal{A}(E_{\alpha a})$ are considered, gives rise to the set

$$\{ H_a^{[M-(M'-\phi)]} \mid M' - \phi = 0, 1, \ldots, M \}$$

of tensors, i.e., precisely the set S_S of Synge tensors associated with $E_{\alpha a}$. In the two-parameter matrix-extensor family of variational integrals, there is an analogous result.

A similar search for normalizing arcs in the class of isoperimetric variational integral pairs

$$\int_{t_1}^{t_2} \sum_{\alpha=\phi}^{M'} \binom{\alpha}{\phi} A_{\alpha a} V^{a(\alpha-\phi)} dt, \int_{t_1}^{t_2} \sum_{\alpha=\phi}^{M'} \binom{\alpha}{\phi} B_{\alpha a} V^{a(\alpha-\phi)} dt = 0 ,$$

$M' = \phi, \phi+1, \ldots, M+\phi$, given by consideration of all $A_{\alpha a}$, $B_{\alpha a}$ in $\mathcal{A}(E_{\alpha a})$ gives rise to the set $S_\lambda = \{ H_a^{[M-R]} | R = 0, 1, \ldots, M\}$ of tensors associated with $A_{\alpha a} + \lambda B_{\alpha a}$, λ a Lagrangian multiplier.

We conclude that, for every tensor T_a in S_H, there is at least one extensor $A_{\alpha a}$ in $\mathcal{A}(E_{\alpha a})$ whose normalizing arcs are described by the vanishing of T_a. Also, for every extensor $A_{\alpha a}$ in $\mathcal{A}(E_{\alpha a})$, the associated Hamiltonian tensor either vanishes identically or is a member of the set S_H.

Another point of interest in the one-parameter case is the result that the normalizing arcs for any extensor $A_{\alpha a}$ in $\mathcal{A}(E_{\alpha a})$ with sufficiently differentiable nonvanishing associated Hamiltonian-type tensor can be described by the vanishing of a Kawaguchi-type extensor of the particular subrealm $\mathcal{M}(M,0)$ containing $E_{\alpha a}$. We recall that, by definition of subrealm of $\mathcal{A}(E_{\alpha a})$, if $A_{\alpha a}$ is an extensor in a subrealm $\mathcal{M}(M',\phi)$ of $\mathcal{A}(E_{\alpha a})$, then there exists a unique set $\{K_L \mid L^* \leq L \leq M\}$ of constants (with $L^* = \max\{0, M - M' + \phi\}$) such that $A_{\alpha a} = \sum_{L=L^*}^{M} K_L(L,M',\phi)_{\alpha a}$. We then have

Theorem (30.3). $A_{\alpha a}\left(= \sum_{L=L^*}^{M} K_L(L,M',\phi)_{\alpha a} \text{ with } L^* = M - M' + \phi\right)$ is an extensor in a subrealm $\mathcal{M}(M',\phi)$ of $\mathcal{A}(E_{\alpha a})$ for which $M - M' + \phi \geq 0$. If $K_{M-M'+\phi} \not\equiv 0$, and if the tensor $H_a^{[L^*]}$, $L^* = M - M' + \phi$, is of class

c^{L^*}, then the normalizing arcs for $A_{\alpha a}$ are given by the vanishing of a pure Kawaguchi-type extensor, namely $(DK)_{\alpha a}[L^*,M,0]$, in the subrealm $\mathfrak{M}(M,0)$ containing $E_{\alpha a}$.

Proof: With $L^* = M - M' + \emptyset$, and by (29.15) with $R = M - M' + \emptyset$, we have

$$(DK)_{L^*a}[L^*,M,0] = \sum_{\mu=0}^{M-L^*} (-1)^{\mu}\binom{L^*+\mu}{L^*} E_{(L^*+\mu)a}{}^{(\mu)}$$

By (30.4) this becomes

$$(DK)_{L^*a}[L^*,M,0] = H_a{}^{[L^*]}$$

By Definition (27.6) and the fact that $H_a[L^*]$ is of class c^{L^*},

$$(DK)_{\alpha a}[L^*,M,0] = \binom{L^*}{\alpha}\left(H_a{}^{[L^*]}\right)^{(L^*-\alpha)}$$

for $\alpha : \emptyset$ to L^*. By (iii) $(DK)_{\alpha a}[L^*,M,0] \equiv 0$ is equivalent to $H_a{}^{[L^*]} = 0$. But by (i) and (ii) we have that the normalizing arcs for $A_{\alpha a}$ are described by $H_a{}^{[L^*]} = 0$, $L^* = M - M' + \emptyset$, and the theorem is proved.

Finally, it is also noteworthy in the one-parameter case that, for any $A_{\alpha a} \in \mathcal{A}(E_{\alpha a})$ (of range $\alpha : \emptyset$ to M' for some permissible fixed

ϕ and M') with nonvanishing associated Hamiltonian-type tensor, the normalizing arcs for $A_{\alpha a}$ are given by equating $A_{\alpha a}$ to some particular linear combination of the extensors $(L,M',\phi)_{\alpha a}$ (which we recall may be regarded as displaced generalized Kawaguchi extensors) of Theorem (26.2).

Theorem (30.4). If (i) $A_{\alpha a}$ is an extensor in a subrealm $\mathfrak{M}(M',\phi)$ of $\mathcal{A}(E_{\alpha a})$ for which $M - M' + \phi \geq 0$, (ii) $K_{M-M'+\phi} \neq 0$, where $\{K_L | M - M' + \phi \leq L \leq M\}$ is the unique set of constants such that $A_{\alpha a} = \sum_{L=M-M'+\phi}^{M} K_L(L,M',\phi)_{\alpha a}$, and (iii) the tensor $H_a^{[M-M'+\phi]}$ is of class $C^{M-M'+\phi}$, then C is a normalizing arc for $A_{\alpha a}$, i.e., along C we have

$$\int_{t_1}^{t_2} \sum_{\alpha=\phi}^{M'} \binom{\alpha}{\phi} A_{\alpha a} V^{a(\alpha-\phi)} dt = 0 \quad ,$$

if the extensor equation

$$(30.6) \qquad A_{\alpha a} = - K_{L^*} \sum_{L=L^*+1}^{M} (-1)^{L-L^*} \binom{L}{M-M'+\phi} (L,M',\phi)_{\alpha a} + \sum_{L=L^*+1}^{M} K_L(L,M',\phi)_{\alpha a}$$

is satisfied along the arc C. Here, as $M - M' + \phi \geq 0$, we have $L^* = M - M' + \phi$.

Proof: First, we notice that the right member of (30.6) is a linear combination of primary basis extensors $(L,M',\phi)_{\alpha a}$ in $\mathfrak{M}(M',\phi)$ for which $L > M - M' + \phi$. Then we note by Theorem (30.1) that each of these primary basis extensors has a vanishing associated Hamiltonian-type tensor. But, as before, the Hamiltonian-type tensor for a linear combination of extensors of the same range is the same linear combination of the associated Hamiltonian-type tensors. Accordingly, we conclude that the Hamiltonian-type tensor associated with the right member of (30.6) vanishes identically, so that any arc C along which (30.6) is satisfied is a normalizing arc for $A_{\alpha a}$.

Alternate proof: We proceed to show that (30.6) is equivalent to the vanishing of the Hamiltonian-type tensor associated with $A_{\alpha a}$, which by (i) and (ii) is the tensor $H_a^{[M-M'+\phi]}$ (given by (30.4)). Now by use of $A_{\alpha a} = \sum_{L=L^*}^{M} K_L(L,M',\phi)_{\alpha a}$ and by subtraction of $\sum_{L=L^*+1}^{M} K_L(L,M',\phi)_{\alpha a}$ from both members of (30.6) we have that (30.6) becomes

$$(30.7) \qquad K_{L^*}(L^*,M',\phi)_{\alpha a} = - K_{L^*} \sum_{L=L^*+1}^{M} (-1)^{L-L^*} \binom{L}{M-M'+\phi} (L,M',\phi)_{\alpha a} \quad .$$

By (ii), $K_{L*} \neq 0$, so that by division of both members of (30.7) by K_{L*} and by collecting terms we have that (30.7) becomes

$$(30.8) \qquad \sum_{L=L*}^{M} (-1)^{L-L*}\binom{L}{M-M'+\emptyset}(L,M',\emptyset)_{\alpha a} = 0 \quad .$$

By Theorem (29.4) the left member of (30.8) is the completely reduced Kawaguchi-type extensor (tensor) $\frac{1}{k_\emptyset}(DK)_{\alpha a}[\emptyset,M',\emptyset]$ with $k_\emptyset \neq 0$ and with α fixed at $\emptyset$. We may then replace (30.8) with $(DK)_{\emptyset a}[\emptyset,M',\emptyset] = 0$, or the equivalent

$$(30.9) \qquad \sum_{L=L*}^{M} (-1)^{L-L*}\binom{L}{L*}E_{La}{}^{(L-L*)} = 0 \quad ,$$

where we have used Theorem (26.2) to replace $(L,M',\emptyset)_{\emptyset a}$ by $E_{La}{}^{(L-L*)}$, $L* = M - M' + \emptyset$. Next, we replace the index L of summation in the left member of (30.9) by μ according to $\mu = L - L*$ to obtain

$$(30.10) \qquad \sum_{\mu=0}^{M-L*} (-1)^{\mu}\binom{L*+\mu}{L*}E_{(L*+\mu)a}{}^{(\mu)} = 0 \quad .$$

By (30.4) we see that this becomes

$$(30.11) \qquad H_a{}^{[M-M'+\emptyset]} = 0 \quad ,$$

or (30.6) is equivalent to the vanishing of $H_a{}^{[M-M'+\emptyset]}$, and the proof is completed.

BIBLIOGRAPHY

1. Craig, H. V., "On Extensors First Order Partial Differential Equations and Poisson Brackets," Tensor, New Series, pp. 159-164, vol. 6, no. 3, December, 1956.

2. ___________, "On an Extended Phase Space," Annali di Matematica, pp. 59-67, Serie IV, Tomo LVIII, 1962.

3. ___________, "On Certain Physical Applications of the Gradient and Kawaguchi Extensors," Tensor, New Series, pp. 212-222, vol. 13, Commemorative Volume I, 1963.

4. ___________, "On Extensors and a Non-Conservative Dynamical System," Tensor, New Series, pp. 144-150, vol. 10, no. 2, May, 1960.

5. ___________, "On Tensors Relative to the Extended Point Transformation," American Journal of Mathematics, pp. 764-774, vol. 59, 1937.

6. ___________, Vector and Tensor Analysis, McGraw-Hill Book Company, New York, 1943.

7. ___________, "On Extensors and the Lagrangian Equations of Motion," Mathematics Magazine, pp. 245-251, vol. 26, 1949.

8. ___________, "On Certain Linear Extensor Equations," Tensor, New Series, pp. 40-50, vol. 4, 1954.

9. ___________, "On Certain Linear Extensor Equations, Paper II," Tensor, New Series, pp. 77-84, vol. 5, 1955.

10. ___________, "On Extensors in the Calculus of Variations," Mathematics Magazine, pp. 175-191, vol. 30, no. 4, 1957.

11. ___________, "On Primary Extensors," Tensor, New Series, pp. 196-206, vol. 8, 1958.

12. ___________, and Horton, C. W., "On Extensors and the Hamiltonian Equations," Tensor, New Series, pp. 47-52, vol. 1, 1951.

13. ___________, "On Multiple Parameter Jacobian Extensors," Tensor, New Series, pp. 27-35, vol. 2, no. 1, 1952.

14. ___________, and Townsend, B. B., "On Certain Metric Extensors," Pacific Journal of Mathematics, pp. 25-46, vol. 3, March, 1953.

15. _____________, "An Extensor Generalization of a Simple Calculus of Variations Problem," Tensor, New Series, vol. 17, no. 3, pp. 313-320, September, 1966.

16. _____________, and Brigman, G. H., "An Extensor Generalization of a Multiple Integral Calculus-of-Variations Problem," Tensor, New Series, vol. 19, no. 1, pp. 69-75, January, 1968.

17. Synge, J. L., "Some Intrinsic and Derived Vectors in a Kawaguchi Space," American Journal of Mathematics, pp. 679-691, vol. 57, 1935.

18. Hombu, H., "On a Non-Finsler Metric Space," Tôhoku Mathematical Journal, pp. 190-198, vol. 37, 1933.

19. Bean, W. C., "Higher Order Euler Equations as Linear Extensor Equations," unpublished master's thesis, The University of Texas, August, 1960.

20. Hildebrand, F. B., Introduction to Numerical Analysis, McGraw-Hill Book Company, Inc., New York, 1956.

21. Gammel, J. L., "A Differentiation Formula," American Mathematical Monthly, pp. 96-99, vol. 57, 1950.

22. Bliss, G. A., Calculus of Variations, pp. 1-40, The Mathematical Association of America, Carus Monograph No. 1, Open Court Publishing Company, LaSalle, Illinois, 1944.

23. Courant, R., "Formalism of Calculus of Variations," Chapter I in Calculus of Variations, Stevens and Company, 1945-1946.

24. Halmos, Paul R., Finite-Dimensional Vector Spaces, Van Nostrand and Company, 1958.

25. Hille, Einar, Analytic Function Theory, pp. 2, 14, 98, vol. 1, Ginn and Company, Dallas, 1959.

26. Gould, H. W., "Combinatorial Identities," West Virginia University, November 12, 1959.

www.ingramcontent.com/pod-product-compliance
Lightning Source LLC
Chambersburg PA
CBHW080531220326
41599CB00032B/6279
* 9 7 8 1 9 3 6 4 1 1 4 0 5 *